MANUELA PERROTTA

BIOMEDICAL INNOVATION IN FERTILITY CARE

Evidence Challenges, Commercialization, and the Market for Hope

BRISTOL
UNIVERSITY
PRESS

First published in Great Britain in 2024 by

Bristol University Press
University of Bristol
1–9 Old Park Hill
Bristol
BS2 8BB
UK
t: +44 (0)117 374 6645
e: bup-info@bristol.ac.uk

Details of international sales and distribution partners are available at
bristoluniversitypress.co.uk

British Library Cataloguing in Publication Data
A catalogue record for this book is available from the British Library

ISBN 978-1-5292-3674-3 hardcover
ISBN 978-1-5292-3676-7 ePub
ISBN 978-1-5292-3675-0 OA PDF

Cover design: blu inc
Front cover image: Stocksy/Liliya Rodnikova
Bristol University Press uses environmentally responsible
print partners.
Printed and bound in Great Britain by CPI Group (UK) Ltd,
Croydon, CR0 4YY

To my mother,
my mother in law,
and my mothers in science

Contents

List of Figures and Tables

Acknowledgements

This book owes its existence to the collective efforts, generosity, and unwavering support of numerous individuals and organizations. Firstly, I am very much grateful to all the participants whose candid sharing of their experiences and insights formed the backbone of this study. While I am unable to name each one individually, their contributions have been indispensable in illuminating the intricate landscape of fertility care.

I am immensely thankful for the support provided by the Wellcome Trust (Investigator Award, grant no. 108577/Z/15/Z) and the British Academy (Innovation Fellowship, grant no. IF2223\230087), which afforded me the resources and time necessary to conduct this research. These opportunities have been instrumental in shaping the trajectory of this work.

Special appreciation goes to Alina Geampana, Josie Hamper, Giulia Zanini, Marcin Smietana, and the entire research team for their invaluable contributions, dedication, and unwavering commitment throughout the duration of this project. Additionally, I am hugely indebted to my project partner Sarah Norcross, Director of the Progress Educational Trust, Sandy Starr, and Jen Willows for their collaboration and support throughout this endeavour.

I am thankful to Elizabeth Mason for her meticulous proofreading and invaluable assistance in refining the clarity of this text. My sincere gratitude also goes to my editor Paul Stevens and assistant editor Ellen Mitchell at Bristol University Press, and my project manager Helen Flitton at Newgen Publishing for their guidance, expertise, and support in bringing this manuscript to fruition. Special thanks to the anonymous reviewers who provided invaluable feedback and constructive criticism on the manuscript, helping to refine and strengthen the final version.

Lastly, my heartfelt thanks to Bruce for his unwavering support and encouragement, and to Carlos for keeping me joyfully motivated.

While the contributions of these individuals and organizations have been invaluable, the responsibility for the interpretations presented in this work rests solely with me.

Introduction: Biomedical Innovation in Fertility Care

On 28 November 2016, BBC One aired a compelling documentary called 'Inside Britain's Fertility Business' (*Panorama*, 2016), cautioning against the growing trend of offering unproven treatments to in vitro fertilization (IVF) patients. The documentary emphasized that numerous costly additional investigations and interventions were being promoted despite a lack of sufficient evidence of their effectiveness, safety, and cost-effectiveness. These treatments were referred to as fertility treatment *add-ons*, a term that gained popularity after the documentary's broadcast, emphasizing their unnecessary nature.

In particular, the BBC documentary drew on the findings of two scientific studies conducted by Oxford University's Centre for Evidence-Based Medicine and published in the *British Medical Journal* (*BMJ*). These studies examined the quality of information provided by clinic websites about add-ons (Spencer et al, 2016) and the quality of the evidence supporting these additional interventions (Heneghan et al, 2016). Both articles expressed significant concerns about the exaggerated benefits that clinics claimed these treatments offered and their unproven nature, meaning they lacked robust scientific evidence.

The issue received significant media attention, sparking heated debates about add-ons in both public and scientific circles. Concerns around the commercialization of fertility treatment and the potential exploitation of patients had already been part of the media's discussions on fertility treatment before the add-on controversy emerged. However, these critics were prompted by the fertility sector's disappointing success rates in comparison to its high costs (Geampana et al, 2018). After the BBC documentary was broadcast, media focus shifted to add-ons, depicting them as unproven treatments that place an unjustifiable financial burden on patients, and private clinics were accused of taking advantage of vulnerable patients. This narrative has remained mostly consistent across various media outlets,[1] despite some subtle variations. However, the media coverage has rarely delved into discussions about what it truly means that a treatment is unproven or how potentially unsafe treatments could be legally offered to patients.

Not only the media, but also the growing medical and scientific literature that discusses add-ons aligns with this representation of these extra treatments. Several publications imply that the scarcity of evidence supporting add-ons is primarily due to the commercialization of the sector, as private clinics have an interest in continuing to sell expensive unproven treatments (van de Wiel et al, 2020; Lensen et al, 2021b; Iacoponi et al, 2022). A minority of voices (Dhont, 2013; Macklon et al, 2019) have raised concerns about the field's ability to produce robust evidence according to the medical criteria in place. However, these concerns are often dismissed as a suspicious defence of the status quo and have not gone beyond the scientific debate.

In the medical literature and public discourse on add-ons, the reason for the popularity of these interventions is usually attributed to the lack of awareness or understanding of the lack of evidence. However, a recent survey (Carrick et al, 2023), in which participants were provided information on the effectiveness and risk of a hypothetical intervention called 'FertiSure', has revealed that a third (34 per cent) of the participants were willing to use it even though they believed it would not improve the probability of having a baby.

In this book, I aim to go beyond the media's simplistic view of add-ons and bridge divides in the debate. I will do so by looking at add-ons as an example of the dominant biomedical innovation model that characterizes fertility care in the UK and worldwide. The books delve into three central topics that underpin the discourse on add-ons: the lack of scientific evidence, the vulnerability experienced by those dealing with infertility, and the effects of the commercialization of fertility provision on biomedical innovation. Despite their significance in the debate, these topics are often oversimplified, failing to capture how complex and interconnected they are. To address this, I approach the subject armed with an interdisciplinary toolbox and draw on wide-ranging polyphonic data obtained from a large qualitative research project[2] conducted in the United Kingdom (UK) and extensive documentary analysis (see the Appendix for further details). Throughout this work, I analyse each issue by focusing on the core uncertainties and unknowns they entail.

The study of uncertainty and 'nonknowledge'[3] has deep roots in the social sciences. Medical sociology has extensively explored how uncertainty is handled in clinical practice and medicine in general (for a review see Mackintosh and Armstrong, 2020), including an extended debate on the epistemological challenges of the introduction of the evidence-based paradigm (Timmermans and Angell, 2001; Timmermans and Kolker, 2004). The field of Science and Technology Studies (STS) has significantly advanced our understanding of how scientific knowledge is generated in practice (Latour and Woolgar, 1979; Knorr-Cetina, 1981; Latour, 1987), and how the inherent uncertainties of knowledge production are addressed in policy-making processes concerning emerging technologies (Frickel et al, 2010; Kastenhofer, 2011; Decoteau and Underman, 2015). The social scientific literature on reproduction has explored at length the uncertainty experienced by those dealing with infertility (Sandelowski, 1987; Franklin, 1997; Becker, 2000), as well as the unfulfilled promises of reproductive medicine to provide certainty in situations of ambiguity (Thompson, 2005; Tjørnhøj-Thomsen, 2005).

Across these bodies of literature, different ontological and epistemological approaches can be found for the analysis of uncertainty. Previous studies (McGoey, 2014) have shown that scientific uncertainty can be exploited by powerful actors for economic gains (such as in the case of the tobacco industry) and political legitimacy. While I include this aspect in my analysis, my interest in this book is to explore these interrelated forms of uncertainties in their generative and performative aspects, rather than viewing them as epistemological deficits. Through the case of add-ons, I explore the uncertainties and unknowns of biomedical innovation to analyse how various stakeholders frame and respond to them. My main argument is that this case underscores the challenges arising by the privatization of healthcare and its unintended consequences. Private fertility clinics embody the tensions between the duties of private businesses operating for profit and their role as providers of essential forms of healthcare not offered by the public service. My core argument is that to understand these tensions and comprehend why fertility patients purchase unproven and often risky biomedical innovations, we need to examine the dynamics of the fertility care market. In this book, I propose the concept of the *hope market* to analyse these dynamics. As I will discuss later in this introduction, I use this term to underscore the specific features of such a market, which is regulated not only by supply and demand through price fluctuation mechanisms, but also by social norms and future expectations.

My approach in this book serves two purposes. Firstly, I aim to move beyond the myopic sensationalism often present in the public discourse on add-ons, as described earlier, and instead, encourage a constructive dialogue among different perspectives rather than fostering conflict. Secondly, I seek to bridge distinct scholarly debates that have emerged in separate academic disciplines.

To fulfil these purposes, the book is organized as follows. In the remainder of this Introduction, I provide an overview of the field of fertility care, its rapid evolution and commercialization, and introduce the concept of the *hope market*. In Chapter One, I challenge the notion of add-ons to underscore that the critical questions this case materializes are inherent in the neoliberal model of biomedical innovation, especially in highly commercialized sectors such as fertility. In Chapter Two, I delve into the notion of 'unproven' treatments, focusing on the uncertainty of medical knowledge and the challenges of generating and interpreting evidence in a highly privatized field with limited standardization. In Chapter Three, I explore the market dynamics and contextualize the allure of novel treatments for patients in the hope market. In Chapter Four, I closely examine the regulatory framework of the UK fertility market and regulators' attempts to address its uncertainties by enhancing patient informed choice. Finally, in the Conclusion, I examine the intricate connections between the challenges of producing reliable evidence, the unintended effects of healthcare privatization, and the limitations in the current regulation of hope markets, identifying key missing elements crucial for fostering responsible biomedical innovation in this field and beyond. Further information regarding the research underpinning this book can be found in the Appendix.

The rapid evolution of IVF

To understand the trajectories of biomedical innovation in fertility care, it is crucial to contextualize it within its historical and recent developments. The birth of Louise Brown in 1978 marked a significant milestone as the first baby conceived through IVF.

While similar efforts were underway worldwide, the first team to successfully achieve a live birth through IVF was based in the UK. The success of the well-known team, comprising Robert Edwards (researcher and physiologist), Jean Purdy (former nurse), and Patrick Steptoe (obstetrician and gynaecologist), took a considerable amount of time to accomplish. The journey began with the successful fertilization of human eggs in Edwards' Cambridge laboratory in 1969, but it took over 100 attempts and nearly a decade to achieve a sustained pregnancy[4] and the birth of a healthy baby.

While the history of early IVF has garnered interest (Johnson, 2011, 2019; Cohen et al, 2015; Elder and Johnson, 2015), very little research has explored its growth in the 1980s. A notable exception is the work of Sarah Franklin (2019), who delves into the transition of IVF from its development within the National Health Service (NHS) to becoming a prime example of health privatization under Thatcherism. As Franklin notes, elected in 1979, just a year after the birth of Louise Brown, Thatcher's political approach had an enormous influence on the development of the field.

When the British NHS was established in 1948, it was designed as a comprehensive and publicly-funded healthcare system, offering universal access to medical services for all UK residents. However, during the period of Margaret Thatcher's policy of privatization between 1979 and 1990, efforts were made to reform the NHS. Although complete privatization of the NHS was not achieved, her policy fostered the growth of private healthcare and the emergence of the so-called two-tier health system.

The British fertility sector was not exempt from the wave of privatization. The Medical Research Council (MRC) and other national funding bodies refused to fund early IVF research (Johnson et al, 2010). When Edwards, Purdy, and Steptoe opened the world's first IVF clinic near Cambridge, they had to rely on an anonymous American donor (Johnson and Elder, 2015). Subsequently, numerous IVF clinics were established, and IVF quickly became a predominantly private and lucrative health service. By 1986, 23 clinics offered IVF in the UK, and this number nearly doubled in the following five years (Franklin, 2019).

Figure 0.1: Number of fertility cycles per year, from 1991 to 2021

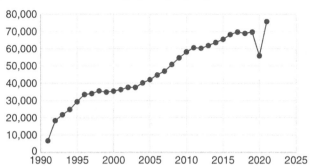

Over the past four decades and beyond, the proliferation of fertility services has been remarkable, resulting in the birth of over 10 million babies through these methods. For example, in the UK, according to recent data published by the Human Fertilisation and Embryology Authority (HFEA, 2023), the UK regulator of the field, the number of treatment cycles has surged from 18,319 in 1992 (the first year with complete data from their registry) to 75,735 in 2021. The significant growth in treatment cycles per year, with the exception of the drop in 2020 due to the pandemic, is illustrated in Figure 0.1. As a result, the number of babies born per year from IVF treatment surged from 1,238 in 1991 to 390,000 thirty years later.

From IVF to fertility care

As the previous section's brief outline indicates, IVF has rapidly transformed from a research and experimental treatment to a normalized routine clinical procedure. In this book, I specifically focus on its clinical translation and routine use, using the term *fertility care* instead of IVF, which is reserved for its medical acceptance.

Technically speaking, IVF is a specific reproductive procedure that involves the collection of eggs and their subsequent fertilization with sperm in a petri dish within a laboratory setting. The resulting embryos are then transferred into the patient's uterus, where they can potentially implant and lead to pregnancy. However, as Sarah Franklin (2013) has illustrated, over the last four decades IVF has evolved into a global technological platform used for various applications beyond its original purpose, including genetic diagnosis, cloning, and stem cell research.

By using the term fertility care, I aim to emphasize two core aspects. Firstly, the distinction between an IVF basic treatment and supplementary treatments or procedures, which is at the centre of the add-on controversy, is not as clear-cut or universally agreed upon as it might seem. As I will illustrate in the following chapters, what is considered 'standard' and 'additional' can vary significantly over time, depending on the circumstances and the location. This book, based on research conducted in the UK, presents a specific local perspective on what constitutes an add-on compared to other treatments or procedures that may be of central concern in other national debates. Moreover, due to the evolving nature of the IVF procedure and the changing opinions on how IVF should be performed, what is considered an add-on at the time of writing might become accepted practice or be rejected in the near future. However, the purpose of this book is not to establish or define what should be considered standard practice. Instead, my aim is to discuss the trajectory of biomedical innovation in the field of fertility care and its implications. By referring to fertility care – which encompasses the entire range of diagnostic testing, medical treatments, and support services aimed at assisting individuals and couples experiencing difficulties in conceiving or carrying a pregnancy – I seek to ensure the resilience of my argument over time.

Secondly, the term fertility care better describes the purpose of the routine use of these treatments, which aim to support individuals struggling with their fertility (one in six people worldwide; see WHO, 2023). While infertility is regarded as a disease, it has been observed that these treatments do not cure it but rather overcome the physical or social[5] obstacles to achieving a pregnancy. Moreover, the indications for undergoing fertility treatments for heterosexual couples have gradually expanded beyond physical conditions, encompassing all causes of infertility, including unexplained infertility. Infertility is, in fact, broadly defined as 'the failure to achieve a pregnancy after 12 months or more of regular unprotected sexual intercourse' (WHO, 2023, pp xi). After this period, potential causes for the inability to conceive should be investigated, but approximately 30 per cent of individuals receive a diagnosis (or lack of diagnosis) of unexplained infertility (Ray et al, 2012). Concerns have been raised regarding the overuse of fertility treatments. For instance, studies comparing couples who had immediate treatment and those who had to wait for treatment show no difference in the number of babies born between these two groups (Carosso et al, 2021). Importantly, while treatment results have improved over the years, fertility care remains a field characterized by low success rates. According to the HFEA (2023), the average live birth rate (LBR) per fresh embryo transfer has risen from 8 per cent in 1991 to 22 per cent in 2021 (27 per cent for frozen embryos).[6] However, it is essential to emphasize that the LBR experiences a notable decrease with age, as depicted in Table 0.1.[7] Additionally, while these figures depict success rates per embryo transfer, they do not encompass cases in which the transfer was impeded due to the absence of available embryos. In fact, there are no available data regarding the cumulative number of live births for women undergoing fertility treatment.

Table 0.1: Live birth rate per patient age group in 2021

Age group	18–34	35–37	38–39	40–42	43–50
LBR	33%	25%	17%	10%	4%

Source: HFEA, *Fertility Trend Report 2021*

These low success rates, together with certain characteristics of fertility care, have significantly influenced policies regarding its funding worldwide. In the next section, I will explore the consequences of the lack of public funding on the commercialization of the field.

The lack of public funding and the commercialization of fertility care

The privatization of fertility care has coincided with a lack of public funding on a global scale. National health providers face the challenge of allocating limited resources across various health services. Specific features of fertility care, such as being viewed as a treatment that does not cure, having low success rates, and occasionally being considered non-essential compared to other medical treatments, have profoundly impacted funding policies worldwide (Dadiya, 2022). Consequently, routine fertility treatment has experienced deprioritization in most countries.

The underfunding of fertility care is not uniform worldwide. In some countries, fertility services receive no funding at all, while others implement eligibility criteria to determine who can access funded treatment. Decision-makers in certain countries use this approach to restrict funding based on social factors like marital status or sexual orientation (Nisker, 2009). In contrast, some countries justify limited funding based on the cost-effectiveness of IVF for specific groups (for an overview, see Luyten et al, 2022), introducing additional clinical and lifestyle criteria, as well as age restrictions, to further restrict access (for a discussion, see Cavaliere and Fletcher, 2022).

These diverse criteria for access lead to a wide range of funding and reimbursement policies and practices worldwide. For example, Israel has one of the most generous public funding policies globally, providing funding for treatment until a family unit has three children, resulting in the highest utilization rate for IVF services in the world. On the other hand, there is no public funding for IVF in the USA, as it is excluded from coverage by federal health insurance programmes and federal research funding. While Israel and the USA represent extremes on this continuum, the overall global funding for IVF is generally inadequate. A survey conducted by the International Federation of Fertility Societies in 2019 revealed that less than half (47 per cent) of reporting countries provided any form of financial support for fertility care, and only 20 per cent of reporting countries offered full reimbursement (IFFS, 2022).

In the UK, the National Institute of Health and Care Excellence (NICE) guidelines (NICE, 2013) recommend that women under 40, who have been trying to conceive for at least two years, should be offered three full cycles of IVF. Women aged 40 to 42 who meet the same criteria and have never had IVF treatment before should be offered one full cycle. It is important to underline that, according to the NICE guidelines, a full cycle should include one episode of ovarian stimulation and the transfer of any resultant fresh and frozen embryos. A full cycle ends when either every available viable embryo has been transferred or one results in a pregnancy.

However, only Scotland has implemented these recommendations, while Wales offers two full cycles and Northern Ireland funds the transfer of one fresh and one frozen embryo only. In contrast, in England, the final decision about who can have NHS-funded IVF in their local area has been delegated to clinical commissioning groups (CCGs), now revised and turned into NHS integrated care boards (ICBs).[8] This has created what is known as the IVF 'postcode lottery' (Wise, 2014), which has raised significant concerns regarding the discriminatory approaches used to determine funding criteria adopted by CCGs/ICBs (for an in-depth discussion, see Tippett, 2023).

A recent audit commissioned in 2021 by the Fertility Network UK, the main patients association in the country, shows that just a tenth (10.1 per cent) of the CCGs/ICBs fund the three full cycles recommended by NICE, while one in five (20.7 per cent) has redefined what an IVF cycle constitutes and offers a partial or reduced cycle. In addition, many CCGs/ICBs have included additional eligibility criteria: almost a quarter (23.3 per cent) do not provide funding for women over 35, and an overwhelming majority exclude from treatment those whose partners have a child from previous relationship.

These variations in approaches and additional restrictions imposed by some CCGs/ICBs significantly impact the total number of NHS-funded cycles across the UK. According to recent data from the HFEA (2023), the number of total cycles funded by the NHS decreased to 27 per cent across all age groups in 2021, down from an average of 40 per cent in the years before 2017. Additionally, the number of NHS-funded cycles dropped to 37 per cent even among patients aged 18–34, compared to 48 per cent of cycles funded for this age group in 2019. An overview of the number of cycles funded privately and via the NHS can be found in Table 0.2.

The diverse policies adopted by the devolved nations and the CCGs/ICBs in England have a significant impact on the geographical distribution of public funding for IVF across regions. As depicted in Table 0.3, the percentage of NHS-funded IVF cycles varies considerably across the UK. In 2021, Scotland had the highest rate of NHS-funded IVF cycles at 58 per cent (down from 62 per cent in 2019), followed by Wales with 30 per cent (down from 39 per cent in 2019), and England with 24 per cent (down from 32 per cent in 2019). Unfortunately, 2021 data for Northern Ireland were not available, leaving uncertainty about any changes from the reported 40 per cent of NHS-funded cycles in 2019.

The shortage of funding and access restrictions have a considerable impact on the experiences of individuals seeking fertility care in the UK. As I investigated elsewhere (Hamper and Perrotta, 2023), patients are acutely aware of the limited availability of NHS funding and the low success rates, which prompts them to consider the option of seeking private care very early in their treatment journey and plan accordingly.

Table 0.2: NHS- and privately-funded IVF cycles in the UK

Year	NHS-funded cycles	NHS funded (%)	Privately-funded cycles	Privately funded (%)	Total
2021	19,963	26.93	54,167	73.07	74,130
2020	15,708	28.52	39,361	71.48	55,069
2019	24,270	34.82	45,426	65.18	69,696
2018	25,500	37.01	43,395	62.99	68,895
2017	27,794	39.87	41,919	60.13	69,713
2016	27,946	40.97	40,262	59.03	68,208
2015	26,728	40.80	38,777	59.20	65,505
2014	26,272	41.29	37,351	58.71	63,623
2013	25,559	41.32	36,303	58.68	61,862
2012	24,126	40.05	36,117	59.95	60,243
2011	24,386	40.26	36,185	59.74	60,571

Note: Data reported in Table 0.2 is the total summary of data from the tab called Table 18 in the underlying dataset available at https://www.hfea.gov.uk/media/5zzh25zw/fertility-treatment-2021-preliminary-trends-and-figures-underlying-dataset.xlsx (last accessed 20 April 2024).

Source: HFEA, *Fertility Trend Report 2021*

In 2021, the LBR across all age groups in the UK was, in fact, 25 per cent (HFEA, 2023). Notably, this rate declines with increasing age, intensifying the sense of urgency in pursuing and purchasing fertility treatments.

It is essential to emphasize that the division between public and private provision in the fertility sector is not as straightforward as commonly presented in public discourse. According to the available data for 2021 (HFEA, 2022b), the UK houses 104 fertility clinics, comprising 60 private clinics and 44 NHS clinics. Notably, the NHS clinics handle 40 per cent of the almost 70,000 treatments performed in 2021, while private clinics carry out the remaining 60 per cent.

Table 0.3: Percentage of NHS-funded cycles across the UK in 2021

Region	NHS-funded cycles (%)
East Midlands	29
East of England	16
London	17
North East	50
North West	37
Northern Ireland*	40
Scotland	58
South East	23
South West	24
Wales	30
West Midlands	28
Yorkshire and the Humber	32

Notes: * Values for Northern Ireland represent 2019; data for Table 0.3 is a re-elaboration of the data underlying the HFEA report on the state of the fertility sector 2020/21, which can be accessed at: https://www.hfea.gov.uk/about-us/publications/research-and-data/state-of-the-fertility-sector-2020-2021/ (last accessed 20 April 2024).

Source: HFEA, *Fertility Trend Report 2021*

However, since the 1980s, there has been a mixture of public and private provisions in the fertility sector, including clinics that are part of private well-known chains but operate on public hospital premises. These 'semi-private' arrangements were enabled by legislation during the wave of privatization aimed at making the NHS more efficient, but they often encouraged the service to be exploited for private financial gain (Franklin, 2019).

These hybrid arrangements create ambiguity. For instance, during an analysis of UK fertility clinic websites (Perrotta et al, 2024), we experienced some difficulty in establishing from its website alone whether a clinic was NHS or private.[9]

Table 0.4: Number of cycles per type of provider and funding

	NHS-funded cycles		Privately-funded cycles		Total
	No. of cycles	%	No. of cycles	%	
NHS clinics	16,436	58.63	11,599	41.37	28,035
Private clinics	7,551	18.44	33,408	81.56	40,959

Source: HFEA, *State of the Fertility Sector Report 2020/21*

Furthermore, it is crucial to note that these hybrid arrangements are reflected in the provision of hybrid public and private treatments: the majority of clinics, whether private or NHS, provide treatment to both NHS- and privately funded patients. These figures are not marginal. As shown in Table 0.4, 41.37 per cent of the cycles performed by NHS clinics are privately funded. Likewise, almost a fifth of the cycles performed by private clinics are funded by the NHS. As a result, patients have often to navigate hybrid treatment pathways leading to varied expectations and experiences (Hamper and Perrotta, 2023).

The scarcity of public funding, coupled with a high level of privatization and the escalating demand for fertility care worldwide, positions this sector as one of the fastest-growing areas in healthcare, poised for exponential growth in the coming years. The global fertility services market was valued at $40.73 billion in 2021 and increased to $47.17 billion in 2022. Projections indicate that it will reach $90.14 billion by 2027, with a compound annual growth rate (CAGR) of 13.6 per cent (IBISWorld, 2022). Currently, Europe holds the largest market share (Precedence Research, 2021), and the UK market size in 2022 was estimated at £542 million (IBISWorld, 2022).[10] As a result, the sector has attracted the attention of private investors (private equity, venture capital, and other financing sources).

In the UK, the recent market expansion has witnessed notable investments from international private equity-backed players, leading to the establishment of new clinics or the acquisition of existing ones. Since 2019, international investors have funded or acquired five of the largest UK clinic networks (Eclipse, 2023).

Recent studies (van de Wiel, 2020a, 2020b; Patrizio et al, 2022; von Schondorf-Gleicher et al, 2022) have explored the financialization of IVF, revealing how private investors have triggered a speculative turn in fertility care. Notably, these studies primarily focus on the USA, where private equity ownership has seen rapid growth across almost all healthcare settings examined (Borsa et al, 2023). A recent review investigating the ownership of fertility clinics in the USA, where just 15 per cent of clinics are situated within academic centres, reveals a transition from private physician-controlled ownership to investor control (Patrizio et al, 2022).

This transition to private equity ownership significantly impacts health outcomes, costs, and quality. A recent review of studies examining the effects of private equity ownership, including two studies related to fertility, has raised concerns about the detrimental impact on costs for patients or payers and the mixed to harmful effects on the quality of care (Borsa et al, 2023).

One of the key characteristics of the private equity investment model is that the firms provide direct managerial oversight to the organizations they acquire, often making changes to increase valuation and future profit potential. As others have highlighted (Hogarth, 2017), the new model of business development in US private equity is centred around a compelling vision for creating value, which is inherently future-oriented and speculative, guided by the concept of disruptive innovation. In the field of fertility care, examples of these new avenues for value creation and market expansion have been identified in the growth of egg freezing (van de Wiel, 2020a, 2020b) and the proliferation of add-ons (Patrizio et al, 2022; von Schondorf-Gleicher et al, 2022) in the USA and worldwide.

In this book, I have chosen to prioritize the term commercialization over privatization or financialization for two main reasons. Firstly, as discussed earlier, the British fertility care sector has been highly privatized since the 1980s. However, the term privatization alone does not fully capture the current changes in the business models adopted. The shift from clinics that are privately owned and managed by medical directors to those owned and managed by non-medical professional managers represents a significant recent change in the sector.

Secondly, in the forthcoming chapters, my focus will be on exploring the intricate network that sustains the commercial practices of IVF clinics in promoting add-ons within a neoliberal biomedical innovation model. While private equity investment has undoubtedly influenced the UK's fertility sector, this book aims to discuss how these commercial practices, driven by this model of business development, have thrived amidst the uncertainties discussed earlier and have shaped the entire sector. Therefore, I use the term commercialization to emphasize that these practices are the result of a complex interplay of factors beyond mere financialization.

Fertility care as a hope market

The analysis of biomedical innovation in fertility care that I propose in this book is rooted in the vast social science literature on infertility and reproductive markets.

The outcomes of IVF's privatization, as Marilyn Strathern has noted (1990), have transformed prospective parents into customers who respond to a market, resulting in the phenomenon of *enterprising up* IVF. Strathern highlights that the market logic in IVF encourages individuals to make market-oriented choices, making it challenging to opt out of consumption.

Despite the global privatization of the fertility sector, there has been surprisingly little scrutiny of its market dimension.[11] Debora Spar's work, focusing on the unregulated North American market, stands as a significant exception. Spar (2006), a Harvard Business School professor and political economist, explicitly argues that reproductive services, such as IVF, surrogacy, and adoption, should be acknowledged and analysed as a market. Her contention is that recognizing the market logics and dynamics could enhance the regulation of the commercial transactions within this industry. As her work emphasizes, this is a unique market, because what is being bought and sold is the promise of a child and, as Charis Thompson highlights, the 'promise is priceless' (2005, p 5). Viewed through an economic lens, this signifies that the desire for these services is not solely dictated by their price. Many individuals deeply committed to having a biologically related child are willing to allocate whatever resources are required, within their financial capacities, to attain successful conception (Spar, 2006). A recent health economics study (Keller et al, 2023) confirms that, unlike in other markets, prices of fertility care have a very limited impact on their demand.

It is crucial to note that the absence of public funding for fertility care and the lack of regulation in terms of treatment expenses further exacerbate inequalities and have significant consequences on patients' lives. For instance, a recent UK-based patient survey (Fertility Network UK, 2023) reveals that nearly half (49 per cent) of participants face financial struggles to cover the cost of their fertility treatment. Many have to dip into their life savings, rely on support from loved ones, and even resort to significant sacrifices, such as selling personal belongings or remortgaging their homes, to pursue their fertility journey. Notably, these market conditions have an even more negative impact on those who cannot afford treatments and experience a double burden: first, from the inability to conceive, and second, from the financial inaccessibility of fertility care (Spar and Harrington, 2009).

As Spar (2010) emphasizes, patients' demands, coupled with the constant influx of information through the internet and new media, sets the stage for a thriving market. In this market, companies introduce new products and optimize profits while giving customers the chance to make purchases in the hope that these products might work for them. While Spar acknowledges that the commodity in the realm of the baby business embodies 'simultaneously hope and medicine' (2006, p 35), neither her work or other analyses of fertility markets delve into the profound impact of hope on shaping these markets.

Extensive social science literature has explored the role of hope in the experiences of IVF patients. In her pioneering study examining the experiences of the initial generation of IVF users in the UK, Sarah Franklin (1997) coins the term *hope technology* to underscore IVF's capacity to provide hope for those struggling with infertility. This emphasis on hope is not solely tied to achieving a successful outcome; instead, it presents IVF as an attractive option even when success is uncertain. While fertility patients are aware of the low success rates of these treatments, they often seek a reproductive resolution to find closure in having attempted treatment, irrespective of whether a biological child is achieved. Similarly, in her study of American couples undergoing IVF, Gay Becker (2000) observes that for people facing involuntary childlessness, fertility care offers numerous treatment options, but limited alternatives outside of it. Alternatives such as adoption or remaining childless are considered only after exhausting medical avenues and depleting financial resources. Both Franklin (1997) and Becker (2000) note that patients need to embrace hope in treatment to persist through failures, safeguarding themselves against potential future regret and contributing to achieving a sense of reproductive closure.

Drawing on these bodies of literature, I introduce the term *hope market* to underscore the influence of hope logics, encompassing aspirations and desires associated with the pursuit of parenthood, on both clinics' provision of and patients' demand for optional interventions in fertility care. Therefore, the following chapters will explore how unproven and potentially risky interventions are marketed and what patients are buying through them.[12] Before delving in this analysis, the following chapter will conduct a comprehensive exploration of these add-ons, drawing insights from perspectives shared by fertility professionals and patients interviewed in this research.

ONE

What Are Fertility Treatment Add-ons?

While fertility treatment add-ons have gained significant attention in medical and public discussions worldwide, the term itself remains somewhat porous and open to interpretation. In the British debate, attention has centred around three main features highlighted in the definition of the term proposed by the Human Fertilisation and Embryology Authority (HFEA, nd).[1] First, they are optional procedures or interventions offered alongside standard fertility treatments, also known as supplementary, adjuvants, or embryology treatments. Second, add-ons often claim to improve the chances of a successful pregnancy or better outcomes, but the evidence supporting these claims for most fertility patients is insufficient or unreliable. Third, add-ons typically involve additional costs on top of the expenses associated with established fertility treatments, varying from a few pounds to thousands.

In the next chapter, I will delve into the controversy surrounding the efficacy and safety of these add-ons, addressing the lack of solid evidence to support the claims made about these interventions. However, in this chapter, my focus will be on critically analysing the concept of add-ons and the diverse perspectives held by various stakeholders, including policy makers, professionals, and patients. Exploring these different interpretations allows us a deeper understanding of the complex landscape and how factors such as cost and specific features influence the definition of what is considered an add-on in different circumstances. Ultimately,

I aim to underlying the ambiguity of the term 'add-on' itself and challenge its usage in the field.

Before delving into this discussion, I will start by providing a brief introduction to the HFEA's implementation of a rating system aimed at providing reliable and transparent information on add-ons. This section also explores how this rating system is perceived by a diverse range of stakeholders. As we shall see, the development of this system has required an iterative process, and despite the rigour of the information it contains, its interpretation by various stakeholders such as patients and professionals differs.

The HFEA add-on rating system

In response to the contentious discussions surrounding the British in vitro fertilization (IVF) sector highlighted in the Introduction, in 2017 the HFEA took a proactive step. They launched a website that aimed to provide patients with information about the available evidence on the effectiveness of commonly used add-ons, while ensuring continued access to these treatments.

The responsibility of assessing the evidence on these add-ons was entrusted to the HFEA's Scientific and Clinical Advances Advisory Committee (SCAAC), along with an independent expert specializing in systematic reviews and evidence assessment. They adopted a conventional evidence-based medicine (EBM) approach, which will be discussed in the next chapter, to categorize add-ons using a rating system. Between 2017 and October 2023, this was referred to as the HFEA 'traffic light' system, as it assigned a colour to each add-on based on the available evidence: green indicates add-ons with multiple high-quality randomized controlled trials (RCTs) demonstrating safety and effectiveness in improving live birth rates (LBRs) for most patients; amber indicates add-ons with conflicting evidence; and red indicates add-ons with no evidence of safety or effectiveness.

Since its introduction, the rating system has undergone periodic reviews, resulting in changes over time. For example, in 2019, the rating for what was previously known as preimplantation genetic screening (PGS), now referred to as pre-implantation genetic testing for aneuploidy (PGT-A) on day five embryos, was changed from amber to red, indicating a lack of evidence for its effectiveness and safety.[2] Moreover, in 2021, the meaning of the traffic light rating itself was altered. Initially, the ratings represented both the effectiveness and safety of each treatment add-on. However, considering feedback from patients and recommendations, it was decided that it would be clearer to use the traffic light system solely for effectiveness. Therefore, additional information on safety has been included on the individual pages of each assessed add-on, while the rating specifically denotes the availability and quality of evidence regarding effectiveness.

In October 2023, further changes were implemented, transforming the traffic light model into a rating based on five categories.[3] This included the introduction of a black rating (indicating sufficient evidence that the add-on has no effect on the treatment outcome) and grey rating (indicating insufficient evidence to assess the effectiveness of the add-on), as well as a further redefinition of the red rating (now indicating potential safety concerns and/or sufficient evidence that the add-on may reduce treatment effectiveness). A summary of the changes in the meaning of the ratings can be found in Table 1.1.

The list of interventions assessed by the rating system has also evolved over the years, as has the rating system, with changes in assessment following the SCAAC's annual evaluation of evidence. Table 1.2 outlines this evolution, including the introduction of new interventions and the removal of others. Notably, changes in the meaning attributed to symbols and colours over the years have impacted each intervention's rating, extending beyond the evidence supporting them. Particularly, safety considerations, previously categorized under 'red', were separated for individual assessment in 2019.

Table 1.1: Meaning of the HFEA add-on rating

2017–2023	
Colour	**Explanation**
Red	We give a red symbol for an add-on where there is *no evidence from RCTs* to show that it is effective at improving the chances of having a baby for most fertility patients.
Amber	We give an amber symbol for an add-on where there is *conflicting evidence from RCTs* to show that an add-on is effective at improving the chances of having a baby for most fertility patients. This means that the evidence is not conclusive and further research is required, and the add-on should not be recommended for routine use.
Green	A green rated add-on has *more than one high quality RCT* which shows that the procedure is effective at improving the chances of having a baby for most fertility patients. These treatment add-ons may be routinely used in fertility treatments and information on these can be found elsewhere on our website, for example the use of intracytoplasmic sperm injection (ICSI) if the cause of infertility is sperm related. Therefore, green rated add-ons will not be included in this review list.
Since October 2023	
Red	*There are potential safety concerns* and/or, on balance, the findings from moderate/high quality evidence show that this *add-on may reduce treatment effectiveness.*
Amber	On balance, it is *not clear whether this add-on is effective at improving the treatment outcome.* This is because there is conflicting moderate/high quality evidence – in some studies the add-on has been found to be effective, but in other studies it has not.
Green	On balance, findings from high quality evidence shows *this add-on is effective at improving the treatment outcome.*
Grey	*We cannot rate the effectiveness of this add-on at improving the treatment outcome as there is insufficient moderate/high quality evidence.*
Black	On balance, the evidence from moderate/high quality evidence shows that *this add-on has no effect on the treatment outcome.*

Table 1.2: List of interventions assessed by the HFEA rating system

Intervention	2017	2020	2023[a]
Artificial egg activation calcium ionophore Chemicals called calcium ionophores are added to the culture media with the aim to initiate a process known as 'egg activation', which is essential for embryo development	Amber	Amber	N/A[b]
Assisted hatching The use of acid, lasers, or other tools to thin or create a hole in the thick protein layer surrounding the embryo with the aim to facilitate its 'hatching' process and aid implantation	Red	Red	Grey
Elective freeze all cycles When all embryos created in a cycle of treatment are frozen for later transfer to the patient's uterus	Amber	Amber	Amber
Endometrial receptivity array (ERA)[c] Tests that claim to determine the optimal timing for transferring an embryo into the uterus to enhance the chances of implantation	N/A	N/A	Red
Endometrial scratching A procedure aimed at injuring the lining of the womb (the endometrium) to stimulate the production of chemicals and hormones, expected to enhance the receptivity of the womb lining for embryo implantation	Amber	Amber	Amber
Hyaluronate enriched medium (e.g. EmbryoGlue) The use of pre-transfer culture medium supplemented with hyaluronan aims to increase the likelihood of successful embryo implantation	Amber	Amber	Amber

(continued)

Table 1.2: List of interventions assessed by the HFEA rating system (continued)

Intervention	2017	2020	2023
Immunological tests and treatments for fertility			
• *Intralipids*	Red	Red	Grey
An intravenously administered mixture of fat and water, intended to supply supplementary nutrients and boost the probability of a successful pregnancy			
• *Intravenous immunoglobulin (IVIG)*	Red	Red	Red
Purified antibodies administered intravenously, intended to enhance the likelihood of a successful pregnancy			
• *Steroids (glucocorticoids)*	Red	Red	Red
A class of drug used to reduce inflammation and suppress immune system activity to enhance the likelihood of a successful pregnancy			
Intracytoplasmic morphologic sperm injection (IMSI)	N/A	Red	Red
A sperm selection method used in intracytoplasmic sperm injection (ICSI), involving the use of a very high magnification microscope to examine detailed images of the sperm			
Intrauterine culture	Amber	Red	Grey
Fertilized eggs are placed in a device inserted into the womb, where they remain for several hours before being transferred to an incubator until they are ready to be transferred back to the womb or used in the future			
Physiological intracytoplasmic sperm injection (PICSI)	N/A	Red	Black
A sperm selection method used in ICSI, which intends to identify the most viable sperm for fertilization using hyaluronic acid			

Table 1.2: List of interventions assessed by the HFEA rating system (continued)

Intervention	2017	2020	2023
Pre-implantation genetic testing for aneuploidy (PGT-A), previously known as pre-implantation genetic screening (PGS)			
• *PGT-A/PGS (day 3)* The biopsy of one or more cells from an embryo on day 3 of development (usually at the 6–8 cell stage) to test for aneuploidy, which is an abnormal number of chromosomes	Red	Red	N/A
• *PGT-A/PGS (day 5)* The biopsy of cells from a blastocyst-stage embryo on day 5 or day 6 of development to test for aneuploidy	Amber	Red	Red
Time-lapse imaging and incubation A range of incubators equipped with integrated cameras continuously capturing images of embryos to assist in selection, with or without the use of algorithms	Amber	Amber	Black

Notes: The accuracy of the table provided is based on the information available at the time of writing. However, changes and updates are anticipated over time. To ensure the most up-to-date and reliable information, please visit the HFEA webpage titled 'Treatment add-ons with limited evidence', available at https://www.hfea.gov.uk/treatments/treatment-add-ons/ (last accessed on 31 January 2024).

a It is crucial to highlight that the latest iteration of the rating system provides additional evaluation per outcome. For example, as of the current assessment, PGT-A is rated red for its potential to increase the chances of having a baby, green for reducing the chances of miscarriage, and grey for the same outcomes in older women.

b In October 2023, artificial egg activation using calcium ionophore was removed from the rated list of treatment add-ons as it should only be offered in specific circumstances.

c In June 2021, the SCAAC approved an application to include the endometrial receptivity array (ERA) in the HFEA's traffic light-rated list of add-ons, which was subsequently added to the website in 2022.

The latest iteration (2023) clearly distinguishes between cases completely lacking evidence (grey) and those showing safety concerns or reduced success rates (red).

This brief overview is crucial for establishing the background necessary to understand the perspectives shared by participants in this research. As their views were gathered between 2018 and 2020, their viewpoints refer specifically to earlier versions of the rating.

Patients' and professionals' views of the traffic light system

The need for reliable and unbiased information in the field of fertility care is unquestionable. The introduction of the traffic light system served as a helpful starting point (see also Lensen et al, 2023) for many in bringing some clarity to a complex landscape. Despite the intended purpose of the traffic light system, its impact has met with mixed results. In this research, a considerable portion of the 51 patients interviewed were unaware that the HFEA website provides detailed information about add-ons specifically for patients. While it is important to note that qualitative research cannot be generalized, a similar pattern was observed in another qualitative study conducted in 2021, which explored the perspectives of patients and professionals regarding the traffic light system in both the UK and Australia (Lensen et al, 2023). Furthermore, according to the latest HFEA patient survey (HFEA, 2022), only 34 per cent of respondents had read the information on the HFEA website regarding treatment add-ons prior to undergoing these treatments.

Patients typically conducted thorough online research, using resources such as the National Health Service (NHS) A–Z directory and various peer support groups. Even among those who were familiar with the traffic light system, the inconclusive nature of the information found on the website contributed to the overall uncertainty surrounding the IVF experience (Silva and Machado, 2010; Haimes, 2013). For example, one patient shares her experience of seeking information about an add-on that has an amber rating:

'I remember looking at the HFEA website and looked at it on there. Probably asked Dr Google, like I do, [laughs] about ten times a day at the moment … I couldn't see anything that particularly led me one way or the other thinking oh, you definitely should do this because it's going to make such a difference or it's a complete waste of time. From what I could see the research kind of seemed to say well, it seems to make sense but we can't really say definitively one way or the other.'

While the traffic light system was regarded as a reliable source of information when compared to other potential sources, the absence of clear-cut guidance on whether to use certain treatments or not generated ambiguity in interpreting the ratings, especially for add-ons categorized as amber. For example, a patient who heavily relies on the website for information provides further insight into this issue:

'I used HFEA website initially on cycle one just really to see is it going to benefit us, what is the benefit, what is the cost. And what do the professionals think about it. But obviously they didn't really have any evidence … and I think that's just from speaking to others and speaking on, you know, your online community and forums. And with most people really saying the same as I am, you know: why would it be there if someone didn't think it might help?'

For many patients who actively seek information, the absence of evidence may engender a sense of possibility that the treatment 'might' potentially be effective. Paradoxically, this intensifies the inherent uncertainty of the infertility journey and reinforces the need to hold onto hope for treatment success. Notably, the excerpt highlights how the assessment and provision of a particular treatment can be perceived as a signal of the potential promise associated with add-ons.

The constant innovation hype that characterizes fertility care often creates high expectations among certain groups of patients, particularly when they are investing substantial amounts in their medical treatments. In this context, the traffic light was perceived by some professionals as a tool to question the emerging equation between innovative laboratory techniques and the quality of treatment provided. Its introduction was particularly welcomed by professionals working in clinics, often within the NHS, where add-ons are either not offered or not explicitly presented as such. Interestingly, some professionals in these clinics justified their non-offering of add-ons to patients by referencing the regulatory stance that clearly states the lack of robust evidence supporting these additional treatments. The marketing of add-ons, especially on clinic websites (van de Wiel et al, 2020; Galiano et al, 2021; Lensen et al, 2021b), often presents them as cutting-edge technologies that are deemed essential for state-of-the-art laboratories. Consequently, some patients perceive clinics without these extra options as more basic. To illustrate, a fertility doctor working in a clinic that did not provide add-ons summarizes this perspective:

> '[The] HFEA has come up with nice information sheet on the HFEA website, the traffic light thing. So it tells patients that, yes, it is not proven which is accepted by the authority. Because previously we tell the patients but they may think okay, this clinic is not doing this, the other clinic is doing this. But somebody needs to tell them that, yeah, this is the evidence, whether X clinic is doing or Y clinic is not doing it, doesn't mean that X or Y are good or bad. Just there's not evidence.'

Furthermore, professionals view this effort to provide clear information as a means to empower patients in making informed choices about their care:

'I think it's about making people aware that, you know, they may choose to do something without evidence being there but actually it's about informed choices. They need to be making informed choices. We don't want people to be going into things paying for lots of stuff because they think if they don't pay for it there's no chance of their IVF treatment working.'

Although the tool's intentions were praised by many, some professionals interviewed expressed concerns about its interpretation. These concerns often revolved around the meanings of the colours (red/green/amber). As others have noted (Lensen et al, 2023), one point raised by several professionals was the absence of green-rated add-ons. Some professionals viewed this absence as casting a negative light on all add-ons, while others believed it to be a consequence of the chosen rating model. For instance, during one of our focus groups, a professional provides an explanation for why green-rated add-ons were not possible:

'If it's green everybody should be offering to everyone. You should be offering it because it's proven to be beneficial so ... that should be your standard treatment. So, you know, they say "Oh, but there's no green on the traffic light system". If it's green, you go, it's no longer an add-on, it's a standard. If there's good compelling evidence that it's beneficial then why would you not be offering it?'

To address the ongoing discussion and clarify that the exclusion of green add-ons from the list was not intended as a negative connotation of add-ons in general, the latest website updates include a further explanation regarding green rated add-ons:

These treatment add-ons may be routinely used in fertility treatments and information on these can be found elsewhere on our website, for example the use of intracytoplasmic sperm injection (ICSI) if the cause of infertility is sperm related. Therefore, green rated add-ons will not be included in this review list. (HFEA, nd)[4]

Performed in the lab by experienced embryologists, ICSI consists of the injection of one single sperm inside an egg, effectively forcing fertilization to occur. While the example of ICSI for severe male infertility is used to address criticism regarding the absence of green add-ons, it is worth noting that the use of ICSI for infertility not due to male factors is not included in the list of add-ons, despite the ongoing global debate on this (as I will discuss further later in this chapter).

The absence of green-rated add-ons in the traffic light system has resulted in various interpretations within the professional community regarding the significance of the amber category, despite the original intentions of the system's creators.[5] Some professionals consider red-rated add-ons as the primary concern due to the lack of evidence supporting their effectiveness. In contrast, they perceive amber-rated add-ons as less problematic, with promising evidence that requires further research to establish their efficacy. However, other professionals raise concerns about the amber category, as it may imply that it is acceptable to charge for these add-ons despite the limited evidence supporting their effectiveness.

The promissory narrative associated with the amber category and the ongoing assessment of add-ons' effectiveness more broadly raise further concerns. As highlighted by one of the professionals interviewed, in the context of limited evidence, it is not possible to foresee whether these interventions are beneficial:

'There's, there was maybe initially an assumption that an amber light meant we were waiting for, as I said before we were waiting for positive, good news to come and not any understanding of the fact that that might not happen and it might be the other way around, we might get the evidence to say it doesn't work, rather than the evidence to say it does work.'

Interestingly, the recent assessment by the HFEA has proven the interviewee's perspective to be accurate. In the latest version of the HFEA rating system implemented in 2023, two previously amber-rated add-ons (physiological intracytoplasmic sperm injection [PICSI] and time-lapse imaging [TLI]) have now been assigned the new black category, signifying sufficient evidence that they make no difference.

Cost of treatment add-ons

One fundamental feature that characterizes add-ons, as discussed earlier, is their association with additional costs, which can vary from a few pounds to substantial amounts. There is considerable variation among clinics in how they price these treatments, and not all clinics directly bill patients for add-ons. This results in a situation where the same treatment may be included in the overall treatment package at some clinics (both NHS and private), while others may charge patients separately for it, leading to significant differences in cost. The unequal distribution of public funding, as explored in the Introduction, further complicates the situation and creates additional scenarios and possibilities. For example, patients undergoing NHS-funded cycles at private clinics could be billed for the add-ons included in their treatment.

Before analysing the influence of costs on the perception and classification of add-ons, it is essential to present an overview of the actual costs and the variations that exist within this landscape. According to the data collected in a systematic review of UK clinic websites regarding the provision of information on TLI, one of the add-ons included in the HFEA rating system, that I conducted with my colleagues in 2022 (Perrotta et al, 2024), we found that the cost of basic treatment varies significantly across clinics, ranging from £3,190 to £7,750, with an average cost of £4,380. As highlighted in a review conducted by the Competition and Market Authority (CMA), comparing treatment prices is challenging due to the diverse ways in which clinics present them, and basic treatment packages across clinics may include different elements (CMA, 2022a).

Additionally, the accuracy of the information available on clinic websites can vary significantly (van de Wiel et al, 2020; Perrotta et al, 2024), and there is limited information available regarding the actual costs incurred by patients for these treatments, making it difficult to determine the total value. Anecdotally, it is known that the average cost of a standard cycle is around £5,000. However, when the costs of add-ons and other services are included, the total cost can escalate significantly, reaching up to £20,000 or more per cycle (CMA, 2022a).

While the complexity of the scenario does not allow for a systematic analysis of the costs of add-ons, I have summarized the prices to patients for some of the most common add-ons in Table 1.3, using data from recent reviews of UK fertility clinic websites (Heneghan et al, 2016; van de Wiel et al, 2020; CMA, 2022a).

These values should be considered as estimates of the actual costs, as there are discrepancies across different reviews due to ongoing changes in the available treatments and clinics' pricing approaches over a relatively short period. As a result, while the table provides some insight into the costs of add-ons, it does not offer a comprehensive overview. The availability of data varies for each add-on included in the traffic light system, and there are numerous other add-ons (which will be discussed in detail in the next section) that are not represented in the table. Additionally, some clinics offer advanced packages that encompass additional treatments such as PGT-A and TLI. These advanced packages can lead to a significant increase in the overall cost, approximately doubling the price of a treatment cycle and reaching up to £9,500 (van de Wiel et al, 2020).

Table 1.3: Estimated costs of most common add-ons

Add-ons	Estimated costs
Assisted hatching	£380–£600
Endometrial receptivity array (ERA)	£450–£1,500
Immunological tests and treatments	£1,270–£2,890
Intracytoplasmic morphologic sperm injection (IMSI)	Up to £1,855
Intrauterine culture	N/A
Physiological intracytoplasmic sperm injection (PICSI)	N/A
Pre-implantation genetic testing for aneuploidy (PGT-A)	£2,100–£3,500
Artificial egg activation calcium ionophore	N/A
Elective freeze all cycles	N/A
Endometrial scratching	£150–£400
Hyaluronate enriched medium (e.g. EmbryoGlue)	Up to £160
Time-lapse imaging	Up to £850

Another example is the EndomeTrio, which includes three different endometrial tests, endometrial receptivity array (ERA), endometrial microbiome metagenomic analysis (EMMA), and analysis of infectious chronic endometritis (ALICE), with costs ranging from £1,450 to £3,100 (CMA, 2022a).

Due to the high costs associated with certain add-ons, the regulator's concern was that some individuals undergoing fertility treatment may exhaust their financial resources after only a few attempts, rather than being able to cover multiple rounds of standard cycles. In recognition of this issue, the HFEA traffic light system website includes the following statement:

If you are paying directly for your own treatment, you may want to think about whether it might be more effective and/or affordable to pay for multiple routine proven treatment cycles, rather than spending large sums of money on a single treatment cycle with treatment add-ons that haven't been proven to be effective at increasing the likelihood of you having a baby. (HFEA, nd)

The influence of cost is also reflected in how both patients and professionals conceptualize add-ons. Although only NHS clinics that offered at least one add-on (TLI) were selected for this research, many of the patients recruited from these clinics were unaware that TLI is categorized as an add-on. TLI is an umbrella term used to refer to all currently available incubators with integrated cameras that allow systematic collection of images of developing embryos. In some cases, patients were not even aware that their embryos were being cultured in this type of incubator. Similarly, when I inquired about the situation with add-ons to a representative from a patient organization based in a Nordic country, her response was simply that they did not have them in her country. This left me questioning whether none of these interventions are used or if they are simply not labelled as add-ons.

According to many of the professionals interviewed, the classification of a treatment as an add-on is solely determined by whether the clinic charges patients directly for it or not. As expressed by a head of laboratory, "an add-on is anything that is charged as an extra on top of the cycle and is an extra line on their invoice. For me that would be a definition of the add-on." However, while this perspective was shared by many professionals, there were exceptions and specific add-ons such as EmbryoGlue (EG) and TLI that highlighted the limitations of this approach. Despite its intriguing name, EG is not a substance that physically binds embryos to the uterus; rather, it is an enriched culture medium used in the laboratory. Similarly, TLI is an incubator equipped with a camera for continuous monitoring and recording of embryo development. Both EG and TLI are rated as amber in the traffic light system and are commonly used, including in NHS clinics. Prior to the introduction of EG and TLI, which are often marketed directly to patients as interventions capable of increasing success rates, IVF laboratories had been using different, less expensive culture media and incubators for years. During my observation at an NHS clinic, an embryologist points out:

'If you are not charging the patient for it, you would not tell them which culture media you are using. We never mentioned the name of the culture media we were using, so we did not think of informing them when we started using EmbryoGlue.'

Many professionals working within NHS clinics stressed that they did not consider these interventions as add-ons because their clinic did not charge patients for them, and often did not even inform them. A trainee embryologist expresses this viewpoint, stating different concerns regarding this perceived lack of advertisement:

'We do the time lapse for everyone, we do the EmbryoGlue for everyone, but it seems to kind of … I feel we should probably show that more to patients, we are doing this; half the time they're not even aware. So we have patients who are like, "please use EmbryoGlue on my embryos" and we're like, we already do, like it's included in the price and they don't really know. I feel that should be advertised more.'

The professional discourse surrounding interventions that are frequently included in the standard treatment provided by both NHS and private clinics reveals the ambiguous nature of the add-on category and its connection to market considerations. Many professionals, particularly those working in clinics that do not charge patients for these interventions, do not necessarily view the use of these treatments without robust evidence as a concern. For instance, this perspective is elucidated by an embryologist:

> 'We do have EmbryoGlue but we don't charge patients for it and we use it for private and for NHS patients because we think it's going to help. If we didn't think it was going to help we wouldn't spend that silly amount of money on it. But, again, if we were to charge for it we should be very careful … Mostly when you have a private unit that it is a company, it's not a charity, it's just a company that does that to make money. It's there to help patients and to help them get pregnant but if it doesn't make money it's just shutting down. And sometimes you are like, oh, I need to keep this job but I don't think this is alright.'

While concerns regarding the charging of patients for unproven treatments were prevalent among the professionals interviewed, only a small number of them questioned the appropriateness of investing significant amounts of public funding in these treatments. As the interviewee emphasizes, the decision to offer a specific intervention at an NHS clinic relies on the internal assessment of the team, which undergoes thorough evaluation to obtain funding for its implementation. For many professionals, the crux of the controversy surrounding add-ons lies, rather, in how they are marketed to patients. The mis-selling of add-ons to patients was a prevalent concern among professionals and often linked to the prevailing notion that patients do not receive sufficient information regarding the lack of evidence supporting the effectiveness of add-ons. This perspective is illustrated by a head of laboratory who stated:

> 'I don't see a problem in using add-ons and I don't see a problem in if you're in a [private] unit and you've got to make money to support it then you're going to have to charge it. But what I do have a problem with is patients

not being fully informed of, and being told that it will, you know, increase their chances by this when it probably, it might not make any difference at all.'

Notably, these examples highlight the complexity and nuances surrounding the category of add-ons, which raises the question that will be addressed in the next section: what should be considered an add-on?

What counts as a treatment add-on for patients and professionals?

Despite their widespread use, there is still ambiguity surrounding what should count as an add-on. The current HFEA rating system includes 13 interventions that are considered the most common add-ons in the UK. These include drugs (such as steroids, intralipid, and intravenous immunoglobulin), genetic tests (for instance, PGT-A), surgical methods (for example, endometrial scratching), and laboratory-based interventions and equipment (including EG and TLI). However, there is no consensus on the exact number of interventions that should be classified as add-ons (Kamath et al, 2019). In principle, any non-essential part of IVF procedures could be considered an add-on, making the category potentially vast.

The recently published guidelines of the European Society of Human Reproduction and Embryology (ESHRE Add-ons working group et al, 2023) contain recommendations for 42 interventions. This list encompasses the add-ons included in the HFEA rating system, along with other interventions that are not typically classified as add-ons in the UK public discourse. Examples of such interventions are screening hysteroscopy, non-invasive PGT, and complementary and alternative therapies (CAT).

The inclusion of CAT sheds light on the ambiguity surrounding the add-on category. CAT itself is a vast category, including acupuncture, homeopathy, nutrition, and reflexology.

Table 1.4: Use of treatment add-ons

Add-ons	2018	2021
Other treatments	29%	35%
Acupuncture	22%	33%
Time-lapse imaging	18%	27%
Hyaluronate enriched medium (e.g. EmbryoGlue)	20%	16%
Endometrial scratch	24%	15%
Immunological tests and treatment	8%	9%
Elective freeze all cycles	11%	9%
Other treatment add-ons	12%	7%
Nutritional therapy, nutritionists and dieticians	N/A	7%
Pre-implantation genetic testing for aneuploidy	7%	7%
Assisted hatching	7%	5%
Endometrial receptivity array	N/A	3%
Intrauterine culture	3%	1%
Artificial egg activation	5%	1%

Note: The data presented in this table have been derived from the HFEA National Patient Survey 2021, which is accessible at https://www.hfea.gov.uk/about-us/publications/research-and-data/national-patient-survey-2021/ (last accessed on 10 July 2023).

As shown in Table 1.4, the latest HFEA patient survey (HFEA, 2022a), revealed that 40 per cent of participants used CAT, with acupuncture being the most commonly chosen add-on, selected by 33 per cent of patients.

Despite its popularity and being considered as add-ons by some fertility professionals (Stein and Harper, 2021), CAT has not been incorporated into the HFEA traffic light system. The HFEA's stance is that they do not regulate CAT as these therapies are typically not offered within fertility clinics. Instead, the HFEA provides a separate information webpage dedicated to CAT, offering general information about what they are and the associated costs. Similar to the add-on webpage, the CAT webpage includes a disclaimer stating that for most patients, routine cycles of proven fertility treatment are effective without the need for CAT.

The pervasiveness of this phenomenon is also not clearly understood. For instance, recent research by van de Wiel et al (2020) has shown that clinical add-ons, such as TLI (67 per cent of clinics), PGT-A (47 per cent), and assisted hatching (28 per cent), are the most commonly advertised by IVF clinics. However, limited information is available regarding other medical and non-medical alternatives that may be offered to patients.

The HFEA patient survey (HFEA, 2022a) highlights that, although the usage of add-ons appears to have decreased compared to the previous survey in 2018, two thirds of participants (65 per cent) still reported using one or more of these interventions. The distribution of add-on usage, as shown in Table 1.4, indicates that those categorized as amber are more popular compared to those marked as red. It is crucial to note that the survey participants were asked to select which add-ons they had used from a predetermined list. As a result, the survey provides limited information regarding the categorization of add-ons itself.

As part of the research upon which this book is based, a patient survey was conducted. Participants were asked if they had or were planning to undergo any additional treatments, without providing a specific definition of add-ons or referring to any predetermined list. The results revealed that 45 per cent of respondents reported including one or more additional treatments or were considering them. By adopting this approach, we collected a comprehensive list comprising over 60 different options, including a range of tests, treatments, and medications. Beyond common interventions such as TLI, EG, and endometrial scratch, these responses encompass a wide range of tests (such as DNA fragmentation test, Chicago test, AMH tests, Hiddec C test), medications (Prednisolone, Clexane, Clomid, Lubion, oral & vaginal Viagra), and other interventions (salpingectomy, hysteroscopy, surgical sperm retrieval, donor egg/sperm), highlighting the array of additional options potentially available to patients.[6]

The ambiguity surrounding the categorization of add-ons becomes apparent through interviews with patients. One interviewee, who underwent treatment at an NHS clinic and expressed strong opposition to add-ons, was surprised to learn that one of the treatments she received, specifically the endometrial scratching procedure ('the scratch'), was classified as an amber add-on according to the traffic light system:

> 'I also think well, if they were add-ons to, that didn't really work and it was just to make money, I don't know if the NHS would be funding them. Mine wasn't to make money and I've been offered them or have, or just had them as standard. So I've not felt that they were add-ons, to me they're just standard things. I've not felt the scratch or anything was an add-on. So I was quite surprised to see it described as an add-on.'

Similar discrepancies regarding what should be considered an add-on are confirmed by other research. For instance, a fertility patient undergoing treatment in an NHS clinic, interviewed as part of research commissioned by the CMA on patients' experiences of buying fertility treatments, considered counselling as an add-on:

> The clinic offered treatment add-ons that would improve chance of success. We opted to go for glue add-on and we did that because it gave you a higher chance of working. We turned down the counselling add-on but maybe would have gone for it if pregnancy was unsuccessful. (CMA, 2022b, p 19)

The ambiguity surrounding the definition and inclusion of add-ons is also evident in interviews with professionals. As previously discussed, many professionals consider the cost borne by patients as the main factor in determining whether a treatment is an add-on. However, this logic does not necessarily apply to all treatments, and specifically to ICSI. As mentioned earlier, ICSI is used as an example of a green-rated add-on for patients with severe male infertility, but there is less consensus on its use for other indications. Due to the conflicting evidence supporting its effectiveness, ESHRE (2023) guidelines do not recommend the use of ICSI for non-male factor infertility. However, clinics often charge higher fees for ICSI due the increase of costs associated with this technique. During the interviews, I asked professionals whether the same principle of cost should be applied to ICSI. Although none of the interviewees had previously considered ICSI an add-on, some shared their views on the matter:

'That's a good point but I would say if you're charging extra then it's an add-on because whether it's IVF or ICSI you're still fertilizing the eggs at the end of the day and I don't think that … Yeah, I don't know. Because, to be honest, with ICSI you are charging, you do charge a bit more for ICSI because it is a bit more of a specialized skill to do ICSI, it takes a lot more equipment and things like that. But I wouldn't call ICSI an add-on so, yeah, that's a good point, I don't know. I'd not really thought about it like that.'

In contrast to some other countries where concerns exist regarding the widespread use of ICSI, in the UK it is considered an established procedure and its usage is generally limited to cases where it is deemed necessary. Consistent with this perspective, many professionals emphasized that the suitability of a particular treatment for a specific group of patients is a crucial criterion for determining whether it should be classified as an add-on. This approach is also endorsed by the HFEA, as stated on the traffic light website:

> For specific patient groups there may be reasons for the use of a treatment add-on other than improving your chances of having a baby. In these situations, it may be appropriate for you to be offered a treatment add-on as part of your treatment and not in a research setting. (HFEA, nd)

Professionals generally agree on the significance of assessing the appropriateness of specific add-ons for particular sub-groups of patients. Some treatments are tailored to address specific underlying causes of infertility in certain patients, but they may not have any impact on those without those particular issues. As a result, some professionals underline that the focus should not solely be on the treatments themselves but rather on their selective and thoughtful use to avoid indiscriminate application. However, the difficulty arises in precisely defining these sub-groups and identifying the patients who would benefit most from each add-on. As highlighted by an embryologist:

> 'I think add-ons probably work for a very small subset of patients. But that patient cohort is so small you'd never be able to select who it would work for … Like EmbryoGlue … EmbryoGlue is meant to help the embryo hatch … There's an enzyme that helps break down the shell of an embryo so that it can hatch out and stick. For most patients you probably won't need it, the embryo will be able to do that itself. There will probably be some other factor that is stopping a pregnancy happening or not helping that pregnancy along. But for maybe, what, like 5 per cent of patients, they may actually need that enzyme to help their embryo to hatch to result in a pregnancy but you'd never know which patients needed it. So some clinics will give it to everyone (a) at an additional cost and (b) to say if you need it then you've had it.'

The embryologist's critique not only highlights the concern regarding charging patients for unnecessary treatments but also introduces a new dimension to the challenges posed by the promissory discourse surrounding add-ons. The precautionary approach, as criticized by the embryologist ("if you need it then you've had it"), fuels an innovation spiral within the field of fertility care, where new treatments are continuously introduced before solid evidence of their effectiveness is established.

Among professionals, there is a broad resistance to the term add-on, and many perceive it to carry a negative connotation acquired within the field. One head nurse succinctly expressed her view, stating, "I hate the term add-ons. I think add-ons refer to things that have very little benefit." As I will delve deeper into in the next chapter, the crux of the matter lies in the absence of solid evidence, which leaves the assessment of benefits and drawbacks to expert opinions, thereby making it controversial to establish what should be accepted as standard practice until high-quality evidence is available.

Challenging the notion of treatment add-ons

As we have seen, the multitude of interpretations and distinctions used to determine what should be considered an add-on highlight the inherent ambiguity of this category. In this final section, I emphasize the risks associated with the ambiguous notion of add-ons, drawing on the concerns expressed by fertility professionals. As discussed in this chapter, add-ons are a residual category, based on what is additional to IVF cycles, which opens up room for interpretations on where to draw the line among the variety of options available to patients to enhance, allegedly, their chances of conceiving. For instance, during a focus group discussion with professionals, a fertility doctor recalls examples of add-ons that patients had inquired about:

'We get asked all the time about like multiple vitamin supplements and acupuncture and all that stuff ... And actually I quite like those conversations because I can say no … you don't need all these supplements. You can actually see them visibly relax and go. But this morning one of my women, she actually said to me, I was on like 20 different supplements for my previous cycle and evening primrose and iron and all that stuff. I said look, your iron levels are 130, it's perfect, you don't need any, you're completely fine, you know. Eat healthily and take folic acid and Vitamin D, that is it. And she was like phew! You could literally see her go back home and I'm just going to simplify my life.'

As this interviewee emphasizes, the ambiguity surrounding the notion of add-ons allows for diverse interpretations, ranging from vitamins and fertility massages to clinical treatments. However, while the former might be considered superfluous and an unnecessary financial burden for patients, they are generally considered safe. As we observed earlier, the add-ons category combines controversial, invasive, and potentially harmful treatments with others that are less contentious. A head of embryology aptly points out that even when considering add-ons offered within fertility clinics exclusively, this category fails to acknowledge the varying degrees of risk associated with different treatments:

'There's a difference between the minimally invasive or a non-invasive technique or procedure and a highly invasive technique or procedure, and I feel very much that a highly invasive technique or procedure almost needs a different set of rules to what is non-invasive. And I think that's a distinction that hasn't been made thus far. So you know, for example, should we be pumping patients full of steroids because we suspect that the natural killer cells are a problem and there's really no evidence for that killer cells are a problem, that's highly invasive, you know. The side effect profile for those steroids is huge, you could cause serious problems … That's versus something where you just, doing something really non-invasive and saying, actually, we think this helps. In which case, other than the potential misleading of patients and charging patients for a procedure that might not be helping them, you're not actually doing any harm.'

The interviewee's remarks not only reinforce the existing concerns about charging patients for unproven treatments but also underscore that the lack of distinction between invasive add-ons that could potentially cause harm and add-ons that may have little impact beyond the financial burden overlooks the potential risks of these treatments. The focus on cost and lack of evidence paradoxically hinders the recognition of some treatments' potential health risks, which could be serious for patients. Although this viewpoint represents a minority, it is significant in questioning the category of add-ons itself and has broader implications for policy and practice.

In summary, the controversy surrounding add-ons has been beneficial in raising awareness about the introduction of innovations before proper testing. However, it has also resulted in a narrow focus on only a few treatments considered add-ons, without challenging the accepted neoliberal model of biomedical innovation. This model promotes the introduction of new interventions without sufficient investigation into their safety and effectiveness, using a revised notion of informed choice to justify their implementation. Throughout this book, my argument is that we should shift our attention to examining this model and its implications, as I will discuss in the forthcoming chapters. Indeed, the central question that remains unanswered in the debate on add-ons is how to distinguish between what is superfluous, dangerous, and what truly provides benefits in the absence of solid evidence. This will be the focus of the next chapter.

TWO

Evidence Challenges in Fertility Care

As we have seen, add-ons often occupy a vexed space in the treatment landscape due to the lack of robust evidence for their effectiveness. In this chapter, I will address the challenges of producing and interpreting evidence in fertility care, particularly within the dominant paradigm of evidence-based medicine (EBM).

The introduction of the EBM movement in the 1990s aimed to standardize medicine by using 'the conscientious, explicit, and judicious use of current best evidence in making decisions about the care of individual patients' (Sackett et al, 1996, p 312). While attempts to standardize medical practice have a long history (Cochrane, 1972; Greenhalgh, 1997; Timmermans and Berg, 2003), EBM emerged with the explicit objective of grounding medical practice on scientific foundations on a large scale by integrating 'the best external evidence with individual clinical expertise and patients' choice' (Sackett et al, 1996, p 312). EBM relies on clinical practice guidelines to disseminate 'proven' medical knowledge, meaning treatments supported by high-quality evidence of effectiveness. However, while EBM has become a hegemonic paradigm in medical practice, it has not fulfilled the expectation of being the ultimate solution for the uncertainty of medical knowledge (see Greenhalgh et al, 2014). Extensive medical and social science literature[1] has shown that producing 'gold standard' evidence to determine whether a treatment is safe and beneficial, and for whom, is a complex undertaking (Timmermans and Berg, 2003; Greenhalgh, 2014).[2] As a result, the promise of EBM to resolve medical uncertainty has not been fully realized. Instead, it has given rise to new forms of uncertainty in medical practice (Timmermans and Angell, 2001).

Firstly, a fundamental challenge in EBM is reaching a consensus on what constitutes high-quality evidence. In the EBM pyramid of evidence (for an overview see Murad et al, 2016), meta-analyses of randomized controlled trials (RCTs) and RCTs are considered top-quality evidence. RCTs have played a significant role in advancing scientific knowledge, particularly in understanding clinical aspects. However, conducting RCTs that produce high-quality evidence requires more than just clinical expertise. It demands advanced methodological and biostatistical skills to ensure the reliability and validity of the findings (Deaton and Cartwright, 2018). The same skills are required to assess the available evidence to determine its quality and reliability, and whether to depend on it (Greenhalgh, 2019). To address the escalating volume of medical evidence, initiatives such as the Cochrane Collaboration have been established. As an independent international organization dedicated to promoting EBM, the Cochrane Collaboration undertakes the crucial task of gathering, evaluating, and summarizing evidence through systematic reviews of healthcare interventions.

While medical evidence has grown over the years, high-quality evidence is perpetually limited. Medical professionals often find themselves in situations where they must make decisions in the absence of conclusive evidence. Sackett and colleagues (1996) argue that in such cases, healthcare providers should rely on the best available external evidence. However, concerns have been raised about using non-randomized studies to assess the safety and effectiveness of new medical treatments (Franklin et al, 2022). Additionally, even when RCTs unequivocally prove the inefficacy of certain interventions that are already in use, these often continue to be used (Tatsioni et al, 2007; Prasad and Cifu, 2015).

In the public and medical discourse of the fertility community, the introduction of new unproven treatments is often seen as a unique characteristic of this field. However, this phenomenon is unfortunately common across various medical sub-disciplines (Epstein and ProPublica, 2017), ranging from knee (Lyu, 2015) and spinal (Bartels, 2009) surgery to the use of statins (Abramson et al, 2013) and stents (Jinkins et al, 2013) for primary prevention of cardiovascular disease. In many cases, these interventions are not entirely useless, but they are overprescribed to individuals who are unlikely to benefit from them (for an extended discussion of the public health implications, see Patashnik et al, 2020). While the lack of evidence is inherent in medical research due to the incremental production of knowledge, concerns have also been raised regarding the potential influence of vested commercial interests within the healthcare industry (Every-Palmer and Howick, 2014; Jureidini and McHenry, 2020; Trayer et al, 2022).

The purpose of this brief introduction is to illustrate that the debates surrounding the introduction of unproven treatments are not exclusive to fertility care but are prevalent across various medical fields. However, it is noteworthy that the social science literature on fertility has largely overlooked this aspect.

In my previous work (Perrotta and Geampana, 2020, 2021; Geampana and Perrotta, 2022; Perrotta and Hamper, 2023), I have extensively examined various aspects of the intersection between EBM and fertility care. While the fertility professionals interviewed in this research rarely openly criticize EBM, the sector as a whole has long resisted its introduction in the field. This resistance is not uncommon across medical sub-fields (Traynor, 2009), as EBM challenges the foundation of traditional professional autonomy (Armstrong, 2002, 2007).

In the following sections, I will delve into this resistance, analysing the challenges associated with generating and interpreting evidence in fertility care. By doing so, I aim to shed light on the complexities and implications of (partially) implementing EBM in the context of fertility treatments.

Evidence-based medicine and fertility care

The encounter between EBM and fertility care needs to be contextualized within the development of in vitro fertilization (IVF). As discussed in the Introduction, being a novel area of medical research, IVF encountered specific obstacles that influenced the integration of EBM principles. During the 1960s, when research was underway for the development of IVF, there were significant ethical concerns about the legitimacy of constructing an evidence base for fertility treatment. While IVF's inception drew from extensive prior research involving animal models, significant apprehensions persisted about the possibility of generating offspring with abnormalities. Moreover, opposition arose due to the inability of potential future offspring to provide consent for the associated risks. Instead, some argued that given the availability of numerous children for adoption, IVF was an unnecessary pursuit. Moral reservations emerged as well, particularly concerning the manipulation of life within laboratory settings and the inevitable loss of embryos during the process. The ethical and moral complexities surrounding IVF led several governments to refrain from offering public support and crucial research funding.

Martin Johnson (2011, 2013, 2019), who has extensively studied the history of reproductive medicine, reveals that despite the high scientific and media attention surrounding the work of Edwards and Steptoe, they faced challenges in securing funding for their scientific and clinical research. Their application, 'Studies on Human Reproduction', did not receive support from the UK Medical Research Council (MRC), which was the major British funder of medical research at the time. Throughout the 1970s, the MRC continued its policy of not funding this type of research. Paradoxically, one of the main hindrances to obtaining funding was that their proposal was deemed purely research-oriented, which exacerbated the ethical and legal concerns mentioned earlier. Only after the successful birth of two healthy babies through IVF did the MRC acknowledge IVF as an experimental treatment, rather than solely a research procedure (Johnson et al, 2010). This reclassification to an experimental treatment helped alleviate ethical concerns, as it was viewed as beneficial for the patients. However, the absence of support from conventional funding sources and the reliance on affluent private donors to finance IVF's development as an experimental treatment posed challenges to the research culture within the field. As Johnson (2013, p 206) notes, this 'probably contributed to the relatively poor performance and the relatively late entry of reproductive medicine to the research evidence base'.

The historical hindering of evidence production in this field has contributed to the current challenges in obtaining high-quality evidence for fertility treatments. A systematic analysis of Cochrane reviews available for fertility treatments[3] conducted by Farquhar and Marjoribanks in 2018 revealed concerning results. They identified 68 Cochrane systematic reviews that reviewed nearly a thousand RCTs (962) related to fertility treatments. Out of the 68 interventions, only a third (23) were identified as effective in the review, with an additional 15 interventions considered promising. However, the majority of interventions (40) were either assessed as ineffective (2) or possibly ineffective (17), or the authors were unable to formulate definitive recommendations (15) due to the absence of sufficient available evidence. Notably, even when conclusions could be drawn, they were primarily based on evidence assessed as moderate quality.

According to the authors, while all the reviews included in their analysis were of high quality, the overall quality of evidence for the various interventions varied significantly. Conducting a frequency analysis of occurrences within the results revealed that the evidence for each intervention was assessed as high quality in only six instances (with two of them being of moderate to high quality). In stark contrast, the evidence was considered to be of low quality in 121 instances, with 39 of those instances falling into the category of very low quality.

This review did not focus on add-ons but rather examined various aspects of fertility treatments. For instance, one area they analysed was whether fertility treatment is indicated for women with unexplained infertility. The review found no conclusive evidence of a difference in live birth rates (LBRs) between IVF and expectant management, which involves periods of unprotected intercourse following completion of fertility investigations.

Unsurprisingly, evidence assessments focused on treatment add-ons have revealed similar issues. The Cochrane special collection specifically focusing on the effectiveness of add-ons in IVF (Lensen, 2020), which collated all the Cochrane reviews evaluating different additional interventions, reveals that none of them are supported by high-quality evidence demonstrating their effectiveness and safety. As in the case of the systematic analysis mentioned previously (Farquhar and Marjoribanks, 2018), the lack of high-quality evidence does not seem to result from the number of conducted RCTs but rather from their average quality. Moreover, in the limitations section of the European Society of Human Reproduction and Embryology (ESHRE, 2023, p 2062) guidelines on add-ons, authors underscore that 'of the 42 recommendations, none could be based on high-quality evidence and only four could be based on moderate-quality evidence, implicating that 95% of the recommendations are supported only by low-quality RCTs, observational data, professional experience, or consensus of the development group'.

A series of articles (Wilkinson et al, 2016; Braakhekke et al, 2017; Stocking et al, 2019; Wilkinson, Brison et al, 2019) authored by a group of fertility researchers involved in evidence reviewing, has raised several concerns regarding the production of evidence in this field. They emphasize that most RCTs suffer from significant limitations. First, there is serious risk of bias due to a lack of blinding, as well as no clear descriptions of randomization procedures and a lack of prospective trial registration. Second, these trials often have recruited too few patients to have enough statistical power to detect clinically relevant effect sizes. As a result, these RCTs are insufficient to determine whether a treatment is beneficial, harmful, or has no effect. Additionally, these trials reported a variety of different outcomes (Wilkinson et al, 2016), making meta-analysis impossible, as I will discuss further in the next section.

The success rates of fertility treatments have witnessed improvement over the past decades, with average pregnancy rates increasing from 10 per cent in 1991 to 29 per cent in 2021 (HFEA, 2023). However, it is important to note that none of the new interventions are anticipated to substantially increase success rates. Instead, their enhancements represent marginal percentages, and this adds complexity to the process of determining their effectiveness through trials.

Additionally, concerns have been raised regarding the trustworthiness of the RCTs and their data integrity. The recent debate on the case of endometrial scratching (ES) illustrates the depth of these concerns. ES is a procedure that involves injuring the lining of the womb (the endometrium) to improve embryo implantation rates. This procedure, rated amber on the Human Fertilisation and Embryology Authority (HFEA) rating system, has been very popular over the last years worldwide (Lensen et al, 2016; Palomba et al, 2023b). Various reviews and meta-analyses (Maged et al, 2023; Palomba et al, 2023a; van Hoogenhuijze et al, 2023), including a Cochrane review (Lensen et al, 2021a), underline that it is unclear whether this procedure increases the chance of implantation and live birth and, if so, for whom, and how the scratch should be performed. Beside the uncertainty regarding its effectiveness, concerns have been raised due to the similar weakness in the quality of evidence both for and against ES. For instance, authors of the Cochrane review (Lensen et al, 2021a) emphasize that while they identified 38 RCTs to be included in the review, many of these were poor quality and at high risk of bias. Therefore, they performed the main analyses only including the eight studies that were not at high risk of bias.

To address these concerns a group of researchers conducted a meta-analysis based on the data collected through different studies (van Hoogenhuijze et al, 2023). After identifying eligible RCTs, they contacted the authors and invited them to share the original data to perform integrity checks and assessed the risk of bias. Out of 52 RCTs identified (including 37 published and 15 unpublished RCTs), less than a third agreed to share data (15 RCTs) over a period of three years, two of which were not included in the analysis after integrity checks. Additionally, of the 13 RCTs (12 published, one unpublished) included in the analysis, ten were assessed as low risk of bias, two raised some concerns and one was identified as high risk.

The concerns related to the dearth of evidence and the integrity of evidence production, as well as the subsequent pleas to improve the quality of RCTs, usually focus on add-ons. However, as this brief overview highlights, the challenges related to the production of evidence in fertility care extend well beyond novel or additional interventions and derive from the accelerated transition of clinical research into routine medical practice.

While the discussion presented in this section highlights the research logic, as noted by others (Dhont, 2013), this seems to be distinct from the logic of medical practice. The following two sections will present and discuss the viewpoints of fertility care professionals actively involved in the daily practice of fertility care. These perspectives will shed light on the challenges concerning evidence production and interpretation, as collected through this research.

Making evidence in a highly commercialized sector

Among the professionals interviewed, some attribute the absence of evidence to inadequate public funding for fertility care, as discussed in the Introduction. One of the fertility doctors emphasizes:

'The point in reproductive medicine which gets ignored, wilfully ignored I would say, is that this is one of the few branches of medicine which is more often practised in the private sector rather than on the NHS. So in the UK only 40 per cent of IVF is done on the NHS and 60 per cent is done in the private sector. And I'm not saying there's anything wrong with the private sector, they provide a good service, but at the end of the day there is less evidence-based medicine because the funding is private, the patient is paying for it.'

As shown in my previous work (Perrotta and Geampana, 2021), the impact of sector commercialization on evidence production is extensive. Firstly, the wide array of products and technologies available introduces numerous potential variations into local laboratory practices, impeding the standardization necessary for conducting RCTs. Additionally, the sense of urgency among patients seeking care (Thompson, 2005), the emotionally and financially taxing experience of infertility (Franklin, 1997), and the associated stigma (Inhorn, 2012) act as barriers to their participation in trials. The challenges of patient recruitment in RCTs are exacerbated by certain characteristics of the interventions under investigation. For example, when discussing research on elective freeze-all cycles, an amber rated add-on according to the traffic light system, an embryologist states:

'That's been really hard to recruit anybody. I think with that study specifically you're asking people to delay their embryo transfer because you want to freeze them and then wait a few weeks. And I think it's really hard to get patients to consent to anything that's going to interfere with their treatment.'

In a standard IVF cycle, typically one to two fresh embryos are transferred, and any remaining suitable embryos are frozen. However, in elective freeze-all cycles, a different approach is taken, where all embryos produced are frozen, and no embryos are transferred during the 'fresh' cycle.[4] As mentioned in the excerpt, this implies that patients may need to wait for a few weeks, or even months, after the hormonal treatment to have their embryos thawed and transferred. Other popular add-ons have posed challenges in recruiting participants for trials with placebo arms, as noted by a fertility doctor, "because everybody wants the add-on as it might be the secret ingredient, and that's the difficulty with research obviously".

Additionally, the lack of funding in this field extends not only to treatments for patients but also to research. A senior researcher specializing in andrology argues:

'It is difficult to convince funding bodies that they [fertility RCTs] are worth funding. In comparison to cancer treatments or care of the elderly that's not seen as a priority ... It's just not seen as something that's worth spending our research pounds or dollars on.'

According to the interviewee, despite infertility being recognized globally as an illness (including by the WHO), funding bodies tend to prioritize other areas of medicine because, as he states, "nobody dies of infertility". The lack of research funding available is seen as a reason for the absence of a research culture, which, in turn, reinforces the problem. To explain the reason for the lack of good quality evidence, a fertility doctor argues:

'There's not much funding. If there is no funding for trials, doing things is a bit difficult. And [a major funding body] gives only one major grant for one trial. And if the previous trials are not done correctly or not finished properly then future trials have a limited success in terms of funding. But when the private companies fund them, it brings an element of bias in the results because they may want to push their own publications or their own products. So unless we have evidence not all questions could be answered. But to answer the questions we need money which is very short. So it's like running around in circles.'

As the interviewee suggests, conducting properly designed and adequately powered trials, rather than relying on small and inconclusive studies, necessitates significant research funding. Poorly conducted trials can make it even more challenging to secure future funding in the field. Moreover, funding is essential to develop the necessary skills and infrastructure to support high-quality research. It is crucial to acknowledge that the principles of EBM must be adapted to the unique context of infertility research, as mentioned earlier. For instance, reaching a consensus on how to report the results of RCTs has been a significant challenge in the field.

The conventional Consolidated Standards of Reporting Trials (CONSORT) guidelines are not sufficient for infertility trials due to their distinct characteristics, involving multiple participants (a patient, a partner, and, if successful, a third individual who is the desired outcome of treatment). To address this issue, a consensus conference was organized in 2013 to modify the CONSORT checklist and enhance the reporting quality of clinical trials testing infertility treatment. This conference determined that the preferred primary outcome for all infertility trials should be live birth (defined as any delivery of a live infant over 20 weeks of gestation) or cumulative live birth, which is defined as the live birth per woman over a specific time period or number of treatment cycles (Harbin Consensus Conference Workshop Group et al, 2014). Despite having this consensus in place for almost a decade, however, the reporting of RCT outcomes remains highly inconsistent, posing significant challenges to establish conclusive evidence, as explained by a fertility doctor involved in evidence reviewing:

'You cannot compare apples and oranges. The data is so heterogeneous then you actually can't pool the results. So if you say that we have three or four trials which are all good but they all reported different things so I can't put it together because you can't, they're not like for like. Then it's pointless, isn't it? … If you cannot run one big trial with 5,000 patients then if you are running ten trials with 500 patients each, then you please all actually report on the same outcomes, then you can pool these together and produce one good result. Go to the top of the pyramid.'

As discussed earlier, this inconsistency in reporting is a primary concern for those reviewing trials. However, as the interviewee underlines, "the only people who are interested in this is people who do meta-analysis". According to another fertility doctor involved with meta-analysis, the lack of active concern regarding improving the quality of evidence lies in the biomedical innovation model:

'[The field of fertility] it's so fast-paced and the margins, again, are so marginal, and the trials, big trials just are more difficult to run. You need many, many thousands of patients to really see a big difference and there's not the big money involved, there's not the money to recruit all these huge numbers of patients needed. The industry is very powerful, and it turns out these technologies that get bought very quickly and start being used very quickly.'

The early commercialization of new treatments indeed plays a pivotal role in the biomedical innovation model (I will delve further into its analysis in the next chapter). However, it is crucial to recognize that the contrasting views between those involved in meta-analysis and those in fertility care should not be trivialized, as they stem from fundamentally different understandings of how knowledge should be produced. These distinct epistemological positions become particularly evident when assessing specific interventions, such as time-lapse imaging (TLI), which comprises a range of incubators with integrated cameras that allow systematic collection of images of developing embryos.

Despite the various advantages TLI offers to professionals (see Perrotta and Geampana, 2020), two Cochrane reviews (Armstrong et al, 2015, 2019) have highlighted that there is no conclusive evidence showing that it can be effective in improving live birth rates. As discussed previously, the analysed RCTs – and the related review – were attempting to assess whether TLI, as an add-on, could increase the chances of success in terms of clinical outcomes. While this logic seems entirely reasonable when evaluating the effectiveness of an add-on before directly marketing it to prospective patients, an embryologist specializing in TLI offers a divergent view:

> 'You don't find people randomized control trialling a new incubator in their lab. They buy an incubator based on cost, based on you know, recommendation. All these different things, they don't buy it because it increases pregnancy rates. So if you look at the TLI for what it is, it's an incubator.'

Looking at TLI as a piece of equipment, specifically an incubator, rather than an add-on, some embryologists with specific interest in technology were questioning the approach to assess such a technology through RCTs. They implicitly suggested that while presenting TLI to patients as an innovative technology capable of increasing their chances of success is problematic; it should not justify treating an incubator as a new form of treatment. Additionally, other embryologists pointed out that TLI itself is an umbrella term encompassing a variety of products, including different incubators offered by various manufacturers, which are distributed with different algorithms to support embryo selection. This diversity in products makes it challenging to establish whether any of these combinations can actually make a difference in terms of clinical outcomes. Consequently, even with the production of new high-quality RCTs, it remains unclear what they actually demonstrate.

For instance, the latest and largest (involving 1,731 patients) RCT on the effectiveness of TLI, published at the time of writing (Kieslinger et al, 2023), showed no significant differences in the 12-month cumulative ongoing pregnancy rate of the group who had their embryos cultured in TLI, with or without the use of algorithms, compared with those incubated in traditional incubators. However, when these results were presented at the annual ESHRE conference, questions were raised regarding whether the results of the RCT pertained only to the model of TLI and related algorithm used in the study (Early Embryo Viability Assessment, known as EEVA) or whether they were generalizable to other TLI models and algorithms.[5] As others have pointed out (Vassena, 2023), the algorithm for embryo selection chosen for the study was up to date when it was designed and approved several years ago. However, since then, technology upgrades have taken place, potentially leading to improved ranking performance compared to the one tested. This consideration raises legitimate concerns regarding the practical value of high-quality evidence, which often requires considerable time to establish. As technology advances, the findings from RCTs may lose some applicability in present-day clinical settings.

In addition, other embryologists have emphasized that, as TLI is a technology, the way in which this is used in different laboratories varies significantly beyond the model of TLI and algorithm employed. For instance, another embryologist specialized in TLI argues:

> 'You can't really do a Cochrane review on TLI … TLI is only a means to look at the embryos but how you select them is not consistent around the world … The way that I see it is that this is not a drug, this is not something that on one arm you have the drug, on the other arm you don't have the drug. I think that's the wrong way to see it … I think that the technology is being assessed without an understanding of how the technology works unfortunately.'

Although this specific critique represents the viewpoint of a minority, this emphasizes how the challenges of producing evidence in this field cannot be solely attributed to the lack of interest or commitment of the professionals working in the sector. As I have highlighted elsewhere (Perrotta and Geampana, 2021), the production of high-quality evidence in this sector requires the standardization and institutionalization of the knowledge production process, which is still ongoing. At the time of writing, conflicting practices and epistemologies are simultaneously active and supported by different groups, making it difficult to establish a shared process.

The challenges discussed in this section are not limited to evidence production alone but also extend to how fertility professionals interpret evidence and handle its absence, as I will delve into in the next section.

Absence of evidence or evidence of absence?

As discussed earlier, while the ongoing debate concerning evidence in fertility care primarily revolves around add-ons, the entire field is burdened by the absence of evidence. Therefore, comprehending how professionals interpret and manage this dearth of evidence regarding the effectiveness of add-ons must be contextualized within the broader environment of a widespread absence of evidence.

As mentioned previously, the absence of evidence in the field finds its roots in the historical development of IVF as a privately-funded experimental treatment. Given the longstanding presence of this evidence gap throughout the sector's evolution, some professionals view the sudden demand for evidence as hypocritical, particularly when it comes from early practitioners who operated without any evidence.[6] Additionally, as we have seen, a significant portion of the procedures included in a standard IVF cycle lack robust evidence-based support. A medical director takes this notion even further, contending that:

'We would never have had IVF in the first place under the current regulatory system because they just wouldn't allow it to happen, because not only is it non-evidence based, it failed for so long that clearly it was never going to work, people believed. So it is, we are in a very difficult situation and I don't think and I don't believe that a firm view that unless it's evidence based it shouldn't be paid for, unless it's evidence based it shouldn't be available is the right way forward. I think we need to be mindful of gaining evidence and we need to be mindful of not misleading patients, but I think we have to be realistic and we have to have the opportunity to try to create the evidence in some way.'

Notably, the case of intracytoplasmic sperm injection (ICSI) was frequently cited to support the argument that the absence of evidence does not imply the evidence of absence of effectiveness. This procedure involves injecting a single sperm directly into an egg and is particularly useful when there are concerns about the sperm's quality (as low sperm count, poor sperm motility, or abnormalities in sperm shape). Since the birth of the first ICSI baby in 1992, ICSI has gained global popularity, transitioning from a technique primarily used for severe male factor infertility to becoming the most commonly used fertilization technique. Recent medical research (De Mouzon et al, 2020) reveals that ICSI is employed in over two-thirds of all assisted reproductive treatments worldwide (68.9 per cent – based on the latest available data from 2012 on non-donor cycles). However, there are notable variations between regions, with ICSI being performed in approximately 55 per cent of cases in Asia, 65 per cent in Europe, 85 per cent in Latin America, and almost 100 per cent of cases in the Middle East.

Although significant declines in sperm count are reported worldwide (Levine et al, 2023), the increase in the use of ICSI is not always justified by an increase of male infertility, as the rising use of ICSI in some regions is not proportional to the number of men diagnosed with infertility. Rather, over the years, ICSI has been widely used to treat both male factor and non-male factor infertility. The success of ICSI is not due to higher pregnancy or live birth rates compared to IVF and there is no evidence of its benefit over conventional IVF in couples without male factor infertility (Haddad et al, 2021; Balli et al, 2022). Therefore, the use of ICSI in non-male factor infertility seems to be unnecessary and remains unexplained. A head of embryology further explains:

'Do you know there's only one publication that is done on a trial to compare ICSI and non-ICSI for unexplained infertility. And they have a N of 27 patients in one and 26 patients in the other. This is the only evidence base we have to suggest that we should have ICSI in the unexplained. I was shocked by that. I mean, this is something which is our bread and butter that we make decisions every day in and day out.'

A Cochrane review (Rumste et al, 2011) that analysed the available evidence on the use of ICSI versus conventional techniques for oocyte insemination during IVF in patients with non-male subfertility found that the evidence (rated as low-quality) was insufficient to draw a conclusion. Recent non-Cochrane reviews (Abbas et al, 2020; Glenn et al, 2021) also confirm that while being more invasive, expansive, and time-consuming, ICSI does not offer any benefits in the treatment of non-male infertility.

Despite the absence of evidence, the ongoing debate in the medical literature regarding its potential overuse (Keating et al, 2019; Quaas, 2021), and the inclusion of ICSI for non-male infertility among the interventions that the ESHRE guidelines on add-ons (2023) do not recommend, the case of ICSI has consistently been presented by several professionals interviewed in this research as a positive example of innovation, highlighting that evidence should not always be a prerequisite. In the UK professional debate, the overuse of ICSI is not perceived as a problem due to its relatively limited use (49.5 per cent of all cycles in 2021, according to the HFEA, 2023). However, this case highlights the extension of the evidence gap and the constant need for professionals to work in this realm of uncertainty, which, in turn, impacts their relationship with evidence itself. When faced with the lack of evidence, they must rely on alternative resources, such as professional experience and expertise, to sustain clinical practice. As a medical director notes, "of course everybody would agree you want high-level evidence-based medicine, but the problem is how do you get it? And therefore, you're in this unknown zone." The next section delves into the discussion of how professionals navigate this unknown zone.

Balancing evidence and the urgency in care

The challenges with evidence production and the generalized absence of evidence discussed earlier also influence how medical professionals conceptualize what logic should be followed in introducing new treatments. As a medical director points out, until conclusive evidence is established you face a professional dilemma:

> 'You have two positions: one is you'll just, you know, consider hopefully highly professional morally practising practitioners offering it to the right people without the full level of evidence being there. It may or may not be beneficial but the patient is being informed and the patient makes their own choice. Or you deny access to what might be a beneficial treatment on the basis that there's no evidence that it's beneficial.'

The latter position is endorsed by both fertility reviewers and a 2019 consensus statement on add-ons (updated in 2023),[7] prompted by the HFEA and signed by ten professional bodies, including the British Fertility Society, Royal College of Obstetricians and Gynaecology, and the European Society of Human Reproduction and Embryology. However, many of the professionals interviewed emphasized how this position is complex to put into practice. One of the doctors interviewed elaborates on this issue:

> 'Some people might say, "When is this super quality RCT going to come? Who is going to pay for it? It might take ten years before this evidence comes out. So until then what are we supposed to do because the patient wants to get pregnant now." So your RCT will take so many years. For funding it takes, to apply takes six months and to get the money two years and to run the RCT, then to publish it. So the patient doesn't want to wait. So what the clinics will say is that we will tell the patients that there is limited evidence, we're not sure, but it's got a reasonable safety profile so if you want you can use it. You think that's right? So there is a lot of debate about this. Some people say what is wrong in doing this, let the patient decide, if they want to pay, what's the problem?'

The complexity of balancing the time needed to establish conclusive evidence and the urgency to care was discussed by many professionals. Some professionals perceived patients' expectation to receive the best currently available care as a driving force for rapid innovation. Discussing a new intervention in use at the National Health Service (NHS) clinic in which they work, a senior embryologist notices that "clinics are under pressure to introduce it straightaway, they're under pressure to improve success rates, and I think that's fine if you're not charging for it".

As discussed in the previous chapter, for most professionals, including those based in NHS clinics, introducing treatment not supported by strong evidence of their benefits becomes controversial when patients are charged for these treatments. However, other NHS-based professionals underline how the pathway to introducing new interventions can be steep when research funding is limited. For instance, discussing the case of ICSI, an NHS research nurse argues:

> 'When I started fertility it [ICSI] wasn't funded on the NHS. If you had male factor infertility the only way you were going to do it was to go in the private sector. The private sector would have used ICSI, the NHS wasn't until they could be shown that it was actually making a benefit and it's cost effective. Well, that's fine but you know, all of those people that had to spend money or go childless because male factor infertility ICSI has been one of the major breakthroughs really in IVF. And so you've got this technology and the NHS isn't going to take it on board until they've got the proof that it works. And quite often it is the private sector that comes up with the innovation first. And the NHS comes in the back of all of their work. Which isn't right but then this is a very, it's an area of medicine that is so poorly funded.'

Acknowledging the need to balance the long time needed to produce evidence with the urge to treat patients now, some professionals suggest basing decisions on the best evidence available, not only RCTs and meta-analysis. For instance, the medical director of an NHS clinic emphasizes that:

'Things are developing so quickly that trying to get a true RCT seems that you're just putting the clock back, because things are already moving further and further down the line. So we have to deal in some cases where there is evidence by smaller studies, by whatever else and have an idea and that will at least ... pilot studies will at least put you in the position to say right, this is more interesting, and at least you can exclude the stuff that doesn't work at that point.'

While this approach is prompted by the generalized scarcity of high-quality evidence discussed earlier and has been adopted for many years, its consequences are often overlooked. Beside the concerns that have been raised in the medical literature (Franklin et al, 2022) on using evidence produced by non-randomized studies to assess the safety and effectiveness of new treatments, in the field of fertility care the cases of some interventions have highlighted the limits of this approach. For instance, the case of preimplantation genetic screening (PGS)/pre-implantation genetic testing for aneuploidy (PGT-A) has been at the centre of the debate as a primary example where professionals' views on an intervention do not change, even after it has been conclusively proven to potentially reduce live birth rates (for a history of the evidence on PGS/PGT-A, see Mastenbroek et al, 2021). In PGS/PGT-A one or more cells are taken from the embryos to screen them for chromosomal abnormalities before they are implanted in the uterus. When this intervention was introduced in the 1990s, it was following the principle that the screening of embryos to check whether they have an abnormal number of chromosomes (known as aneuploidy) would avoid the transfer of embryos that do not have potential for life. While early studies showed some improvements in pregnancy rates for embryo transfer, subsequent RCTs showed decreased chances of ongoing pregnancy in comparison with IVF without PGS. PGT-A, the new genetic testing performed on day five with a different technique and analysis than PGS, is currently rated red on the HFEA rating system. As an embryologist explained in one of the focus groups:

> 'Nobody needs to do genetic screening because it was, you know, really inaccurate and it just wasn't worth it. It stopped for years, it's now come back, a lot more people offer it but now it's just been turned red on the HFEA add-on thing, like in the last few days, and we were all kind of thinking we're not too sure about PGS, it's still quite invasive but you just, you don't know.'

However, other participants were supportive of PGT-A and suggested that the purpose of this technique is not to improve success rates but has other advantages. For instance, these advantages were explained by a medical director of a clinic that offered PGT-A for an additional cost to some patients:

> 'If they've had a current pregnancy loss, if they're older and expecting a good embryo yield, we discuss the potential of the advantages of a shortened time to pregnancy and avoidance of multiple frozen cycles which may be inefficient. For private patients they can always pay for it anyway if they want, but we would only, we would certainly suggest that patients who are going to put a large number of embryos into a freezer … It avoids them paying for treatment that is destined to fail. So in those respects it's either cost-neutral or cost-beneficial to do it so it's actually reducing their costs … And I'd tell the NHS patients it is not going to change your pregnancy rate. If you put eight embryos in the freezer the NHS is paying for each of those embryos, um, at present the

NHS will pay for each of those embryos to go back and you will get pregnant. All you're paying for is a shortened time to pregnancy and the removal of aneuploid embryos that couldn't get you pregnant.'

As indicated on the HFEA rating website, which aligns with what is suggested by this professional, PGT-A is usually offered to women over 37, patients who have had several miscarriages or failed IVF cycles, people with a family history of chromosome problems, and men whose sperm may carry abnormal chromosomes. While PGT-A's potential to reduce the chances of miscarriages is confirmed by the current HFEA assessment (rated green for this specific outcome), PGT-A is overall rated red to signal that there is no conclusive evidence supporting its effectiveness in enhancing success rates. The potential benefits gained from embryo selection are cancelled but the reduction in available embryos for transfer after PGT-A. Moreover, within the same embryo, cells can possess varying genetic compositions, a phenomenon known as 'mosaicism'. This complexity sometimes causes PGT-A results to indicate chromosome problems in embryos that could still result in healthy pregnancies, or vice versa.

These concerns have grown more prominent in the medical community after the publication of a recent clinical trial (Capalbo et al, 2021), which was published after the data collection period of this research. The trial shows comparable live birth and miscarriage rates among embryos exhibiting no abnormalities, as well as embryos with a low or moderate proportion of abnormal cells (referred to as low-grade and middle-grade mosaic). These results suggest that many embryos previously discarded due to genetic testing might have had the potential to lead to pregnancies and healthy babies. Unfortunately, these possibilities were lost due to the application of this technique.

The case of PGS/PGT-A exemplifies the risks associated with adopting new interventions before thorough assessments have been conducted. These interventions not only might lack benefits but could potentially lead to harm as well.

What do patients think of evidence?

The scientific literature (Lensen et al, 2021b), the professional discourse, and the regulatory approaches (explored in the last chapter) attribute patients' decisions to opt for unproven or risky treatments to inadequate information or a lack of understanding of what constitutes solid evidence. In turn, this lack of understanding is associated with the poor quality of information within the sector, including content found on clinic websites (van de Wiel et al, 2020; Galiano et al, 2021; Perrotta et al, 2024). Interestingly, despite the heated debate on the lack of evidence supporting add-ons and the strong assumption that patients may either lack awareness of the insufficient evidence or struggle to understand what constitutes robust evidence, little is known about how fertility patients interpret evidence and the role this has in their decisions about their treatment. While transparent and reliable information is crucial for patients to take informed choices about what to include in their treatment, interviews with patients in this research show more complex scenarios.

As I have discussed in detail elsewhere (Perrotta and Hamper, 2023), overall, the patients interviewed interpreted medical evidence and its role in their decision-making regarding treatment in very different ways. It is crucial to emphasize that patients are a heterogeneous group with different backgrounds, needs, and priorities. Unsurprisingly, in this research patients' approaches to evidence ranged from those who disengaged from questions of evidence to those who took a proactive stance in their assessment. These differences were often based on the biography of participants, their personal circumstances and the stage of their infertility journey. While a systematization of these approaches can be found in my previous work (Perrotta and Hamper, 2023),[8] here I focus on the two contrasting standpoints that emerge from patients' interpretation of evidence.

Patients that actively disengaged with evidence evaluation emphasized how they wanted to trust their doctors. While this standpoint could be interpreted as a lack of understanding of evidence, their interviews highlight that often these patients intentionally chose to delegate evaluations of evidence to medical professionals and did not want to take on the responsibility for deciding about their treatment. On the contrary, other patients who were considering using unproven treatments, and were often actively seeking them, were highly involved in assessing evidence as an attempt to direct their treatment. While I will explore this standpoint and its implication in the next chapter, in this section I focus on those who were willing to trust their doctors.

It is essential to notice that, most often, patients start fertility treatment after a long period of trying to conceive naturally. Being finally in the hands of experts can reduce the uncertainty of infertility, as this patient emphasizes:

'Once I was in the IVF process there was a sense of … I can relax a little bit, I am with experts and with the top people. There aren't, you know, anybody else who knows more than this, than these people, and I could put my faith and my trust in them. And I think I did almost completely.'

Some patients underline how, not being medical experts themselves, trusting their doctors' recommendations is the best possible option for them. For instance, a patient discussing taking a medication suggested by their clinic comments:

'I hadn't really looked into it [the medication] myself. I'm just happy to go along with what they suggest because they really know what they're talking about so when he offered it me I was like, well, it can't hurt. Yeah. So yeah, just trying anything they offer me really.'

While most patients who are willing to trust their doctors claim a need to put themselves in the hands of experts, their confidence in fertility clinics is often dented by concerns regarding potential vested interests. For instance, acknowledging an understanding of the lack of evidence supporting treatments, a patient explains:

'I think there does need to be probably a lot more, you know, a lot of backup statistics but I think the doctor, I'd like to think or I felt anyway, knew his stuff, you know, enough to say yes, we'll try this or no, don't bother to try that, that's not going to work for you. I don't think it's going to work but if you really want to do it then obviously you can do it ... I think the doctor knows your body and the issues enough to hopefully, to hopefully offer an add-on that he thinks would actually be beneficial and not just because it's money-making.'

This need to trust doctors while taking into account the possibility that their recommendations might suffer from some financial influence was addressed by patients in different ways. For instance, a patient notes that being treated by the NHS is reassuring as it avoids these concerns:

'I suppose that's what makes me feel so comfortable about having something done on the NHS ... They are only doing it if they think it's definitely going to work. Money is tight, resources are tight, we're just going to do the stuff that we know works really well.'

As other patients note, there is no reason not to trust doctors' judgments if treatments are not charged for. For instance, a patient undergoing NHS-funded treatment at a private clinic that included various unproven interventions, including TLI, EG, and ES, explains:

'I think the scratch might be like an amber now, I think it was downgraded or upgraded recently but again because they're included for me so there's no real reason not to do it. So you know, I'm going to carry on doing them if it's there, so. If it came to where I had to pay for them then I might look more into the evidence.'

On the contrary, patients who have to self-fund their treatment often rely on their perception of not being pressured into buying additional treatments to trust their doctors. For example, a patient who had exhausted their funded cycles and was undergoing privately-funded treatment at one of the private clinics that was part of one of the largest chain of clinics in the UK explains that while they are aware of the controversy regarding some of the interventions offered, they would return to the same clinic. To explain their decision the patient adds:

'I think there's, there is some things that clinics try and get extra money from the weak or the vulnerable or however you perceive it. But my experience wasn't that. My experience was people who really want you to have a baby and that's, we put our trust in them.'

It is crucial to acknowledge that these standpoints are not mutually exclusive and can change over time. In particular, often patients become increasingly involved in directing their treatment if they have experienced several unsuccessful attempts. Additionally, sometimes patients' standpoints are opposite within the same couple, which can add to the burden of fertility treatment as this patient explains:

'I think I found it quite stressful. I found it really stressful actually. That I just felt, I felt quite on my own with it, I felt quite like … My husband is great, he's not medical at all so he just was, he's very much a person who's just, I'll just do what I'm told by the doctor, you know, that is his approach. So I felt that, you know, I'm not really like that and then I thought well, I need to look into these things.'

In this section, I have delved into the intricate dynamics of trust patients place in their doctors. In the next chapter, I will narrow my focus to individuals such as this last patient, who find themselves compelled to take a more active role in shaping their treatment journey.

THREE

The Fertility Market:
Help, Hype, and Hope

In view of the uncertainties around the potential benefits or risks of some fertility interventions, a lingering question remains: who holds sway, clinics or patients? According to the opponents of these treatments, these are often promoted by private clinics driven by financial gain. In contrast, promoters defend patients' choices, which are based on their needs and priorities.

Among the professionals who participated in this research, a common view was that some patients ask for these treatments. For instance, while engaging with a fertility doctor at a social gathering during a Fertility Show,[1] he pointed out that "patients on Mumsnet [an online platform][2] are shaping the market in the UK as they tell stories about pregnancies". According to this senior doctor, who was based in a renowned London private clinic, as patients share their experiences on this platform, anecdotal accounts become evidence that a certain treatment works. To exemplify this, the doctor continues: "If a popular blogger says that she had three unsuccessful cycles and then she got pregnant in a fourth cycle that included time-lapse imaging [TLI], you will have a group of patients that will ask for that treatment. If you do not offer it, you might lose clients. So you might decide to have this tool not because it works, but because you are afraid of losing clients."

This commercial perspective was often criticized among the professionals interviewed. However, the focus on the commercial side of fertility care, which explicitly refers to the loss of clients in a fertility market, is crucial to unveil some of the dynamics that determine the adoption of novel interventions in this field. Additionally, most professionals, including those working in National Health Service (NHS) clinics, acknowledged the pressure they feel to provide the latest interventions to meet patients' demands. For instance, a senior andrologist working in an NHS clinic points out:

'Dr Google has a lot to answer for. So patients will go and Google and then they'll read all the claims on the internet and then they will come in with a shopping list of what they think they want. And I think that makes the job of doctors and nurses very difficult, because you're then having to tell people why they shouldn't do something. And I think that, for people in the NHS, that makes potentially for a bad experience of the NHS, because the NHS isn't going to fund something if they don't think it's going to work. But if you didn't get pregnant and you didn't have this [intervention], you're going to blame it on the fact you didn't have it. And then you go and leap into the private sector, where you ask for it and you get it and you get pregnant. So you're always going to say that it was the thing that you bought that made you pregnant. But what people don't know is that if they'd stayed in the NHS and just had another cycle, they might have got pregnant the second cycle without it. So it becomes self-fulfilling, it becomes self-perpetuating. And I think it puts health professionals in a really difficult spot when they have to say no.'

In the next section, I will discuss "all the claims on the internet" that this participant is referring to, as they come from a variety of sources. Before delving into that, it is crucial to note that, although patient-driven pressure was very common among professionals interviewed in both the NHS and the private sector in this research, a recent UK patient survey (Cirkovic et al, 2023) indicates an opposite trend. Contrary to the professionals' experiences, patients most often report being offered add-ons by their clinics. Among the participants in this survey, 81 per cent of those who opted to use one of these treatments claimed that it was suggested by their doctors, while only 19 per cent were requested by the patients themselves. Private patients also tended to ask for add-ons more often (47 per cent compared to 29 per cent for NHS patients) and use them more frequently (74 per cent compared to 52 per cent for NHS patients).

Determining whether these interventions are promoted by clinics or sought by patients poses several challenges. Recent qualitative studies (Armstrong et al, 2023) exploring views on add-ons in the UK and Australia report similar contrasting perspectives between clinics and patients. Both patients' and professionals' views are based on their perceptions and reciprocal intentions might be misinterpreted. Additionally, the findings of a similar Australian survey (Lensen et al, 2021c) reveals that treatment decisions are commonly shared. Most patients who participated in this survey felt that they shared the decision with their doctors equally (51 per cent, with a range from 40 per cent to 85 per cent), a fifth (18 per cent) thought doctors had no say, while others (7 per cent) thought doctors had all the say. Therefore, simplifying the decision to include unproven and potentially risky interventions into binary terms might not accurately capture the complexity of treatment decision-making.

In this chapter, instead of debating whether the use of these interventions is propelled by clinics or patients, I will examine the convergence of clinics' provision and patients' demand for these interventions in the fertility market.

Selling biomedical innovation

The social scientific literature (Franklin, 1997; Hawkins, 2013) has explored how, in order to showcase patients' happy ending stories, fertility clinic websites most often use pictures of 'dream' or 'miracle' babies born through the use of cutting-edge biomedical technology. Sarah Franklin (1997) notices how these messages emphasize the potential of these technologies and the faith in scientific progress. More recent analyses of clinic websites have shown how similar messages characterized the innovation rhetoric surrounding many of the commonly used interventions discussed earlier (Swoboda, 2015; Takhar and Pemberton, 2020).

A prime case of this innovation rhetoric is depicted in the presentation of TLI on UK fertility clinics websites. A systematic review of UK clinic websites regarding the provision of information on TLI that I conducted with my colleagues in 2022 (Perrotta et al, 2024) reveals that TLI is primarily described as a more 'advanced', cutting-edge incubator technology. The narrative presenting TLI as ground-breaking piece of laboratory equipment is very common, independently on whether clinics charge patients for TLI or not. Most websites (90 per cent) emphasize that, thanks to the additional information provided, TLI helps embryologists to select the embryo that is most likely to result in a successful pregnancy and/or live birth. While many professionals interviewed agreed on the potential of TLI (for details, see Perrotta and Geampana, 2020), the benefits emphasized on fertility websites lack strong supporting evidence (Armstrong et al, 2015, 2019) and recent evidence assessments contradict these claims. In October 2023, its Human Fertilisation and Embryology Authority (HFEA) rating (previously amber) was designated as black, indicating sufficient evidence suggesting no significant impact on pregnancy outcomes. Furthermore, the European Society of Human Reproduction and Embryology (ESHRE, 2023) guidelines do not recommend its use to enhance success rates.

Interestingly, while a growing literature has scrutinized the information provided on fertility clinic websites (Spencer et al, 2016; van de Wiel et al, 2020; Galiano et al, 2021), little attention has been paid to the companies that manufacture interventions that are laboratory equipment or biomedical devices. These include, for instance, TLI and EmbryoGlue (EG), two of the most used additional interventions in the UK and globally. Although these products are acquired and used by fertility clinics, the marketing of manufacturing companies often targets fertility patients directly. For example, Vitrolife, a leading company specializing in fertility care products, including EG and Embryoscope, the most widely used brand of TLI, has a dedicated section on its corporate website that showcases these products to patients.[3] This section of the website, evocatively named 'the IVF journey', describes all the products Vitrolife offer to 'maximise success every step of the way'. The innovation narrative fostered in this website underscores their contribution to the scientific progress in this field. As the website emphasizes:

> Vitrolife has been at the forefront of IVF since the beginning, with a 30-year track record of development and collaboration with universities, clinics, and customers worldwide. The result is an unbroken chain of innovative high quality products that ensures optimal care at every step throughout IVF treatment.

Furthermore, the Vitrolife webpage[4] presenting their TLI branded models to patients emphasizes that TLI should be standard care as 'time-lapse technologies by Vitrolife have been shown in multiple clinical trials to improve clinical outcomes'. Interestingly, the dedicated Vitrolife webpage[5] where the evidence supporting the benefits of their TLI technologies are presented does not engage with a discussion on how (or whether) TLI can improve clinical outcomes. Rather, it emphasizes several other benefits, including documented improved embryo development, optimized use of clinical resources and better patient communication. In particular, while referring to the documented improved embryo development (including seven references of articles published on well-known fertility scientific journals), it is noted that 'this *may lead* to an improved clinical outcome' (emphasis added). However, there is no additional emphasis on potential improvements to clinical outcomes or specific claims of what these could be (for instance, pregnancy or live birth rates or reducing time to live birth or miscarriages).

While the website indirectly promotes the message that TLI has the *potential* to improve success rates without engaging in a detailed discussion of evidence, the parallel webpage[6] presenting EG heavily relies on this discussion to promote its effectiveness. The website discusses how EG can 'promote implantation and increase pregnancy rates', including a section titled 'The proven effectiveness of EmbryoGlue' that quotes directly from the most recent Cochrane review (Heymann et al, 2020). In this case, the ambiguity generated by the lack of evidence is reflected in different interpretations. As correctly quoted on the website, the Cochrane review concludes that 'moderate-quality evidence shows improved clinical pregnancy and live birth rates with the addition of hyaluronic acid as an adherence compound in embryo transfer media in ART'. However, while the Vitrolife website presents the results of the Cochrane review as conclusive evidence, in the results section of the review the authors states that 'adding hyaluronic acid to transfer media *probably* results in an increase in both clinical pregnancy and multiple pregnancy rates' (emphasis added). Interestingly, the case of EG shows how the same evidence base can be interpreted differently following different criteria. While a positive assessment has been confirmed by the ESHRE guidelines, which recommend the use hyaluronic acid addition to transfer media, the HFEA has adopted a stricter evaluation of the same evidence-base, rating it amber. Notably, the HFEA's Scientific and Clinical Advances Advisory Committee (SCAAC) deliberated that, despite the promising benefits, additional significant RCTs would be necessary before assigning a green rating to EG, given the current (moderate) *quality of evidence*. Beyond quoting the Cochrane review results, the Vitrolife website does not provide any discussion of the quality of available evidence. Rather, drawing on the results of the Cochrane review, the Vitrolife webpage states:

> The live birth rate increased from 33.3% to 40.2% with the use of EmbryoGlue. According to a number needed to treat (NNT) calculation, based on the Cochrane Review, one additional live birth was achieved for every 14 transfers. For a clinic with 700 cycles annually, this is one additional baby born per week, or *a 20% increase in live births* [emphasis added].

While the claim of a 20 per cent increase in birth rate, calculated over the initial 33 per cent success rate, is mathematically accurate, it serves as a prime example of the *hype* promoted in this sector and could be easily misinterpreted by a non-expert audience. The past-Chairman of ESHRE, Evers (2017) has highlighted that the hype surrounding these interventions is due to their inflated expectations, which, due to the lack of evidence, remain unconfirmed or debunked. It is important to emphasize that, as shown by STS scholarship, technological hypes are common both within and outside biomedical innovation processes (Brown, 2003; Borup et al, 2006; Ruef and Markard, 2010). However, they are not neutral; rather, they have a perfomative nature. They provide legitimacy, help to attract financing, and shape the expectations of the actors involved, thus shaping the dynamic of the innovation trajectories of these interventions (Van Lente et al, 2013). In the case of add-ons, the hype contributes to their proliferation. As others have noted (Carson et al, 2021), the emphasis on progress and potentiality that permeates the narratives of fertility clinics and product manufacturers might have an impact on patient–provider relationships, influencing patient autonomy and treatment decision-making. As I will discuss in the following sections, these innovation narratives shape patients' experience and influence both their treatment decision-making and their decision to purchase certain interventions.

Contextualizing treatment decision-making in the hope market

Before delving into the analysis of the reasons that prompt patients to contemplate and frequently pursue these additional interventions, it is crucial to provide context for the decision-making process regarding the use and financing of these interventions within the broader scope of treatment choices. Importantly, as highlighted by others (Strathern, 1990), the process of making treatment decisions is intertwined with a broader anticipation of *the need to make choices* related to infertility. A patient who has undergone various self-funded treatment cycles, including the most recent one at an NHS clinic, elaborates:

> 'Going for IVF it's like ... You obviously grow up with like, "I want to have my baby, I want to ..." Like it's not what you want, right? I didn't choose IVF, so it's quite a lot to get your head round, you know. Oh, I think it's going, maybe it will still happen. It's like me and my husband, we're not like, we were like, well, let's do this, takes a while for us to make decisions sometimes. But we'll make them in the end.'

The decision this patient is referring to extends beyond the medical aspects of treatment. They encompass choices like initiating the procedure instead of waiting, selecting a clinic, and timing treatment cycles in alignment with other life commitments. For this particular patient, after an unsuccessful NHS-funded cycle, subsequent decisions revolved around opting for a private clinic or revisiting an NHS clinic for a privately-funded cycle.

Similarly, patients' decision to opt for unproven and potentially risky treatments can only be understood in the context of their experience of infertility. As discussed in the Introduction, early studies of in vitro fertilization (IVF) users (Franklin, 1997; Becker, 2000) reveal that patients feel compelled to explore all options before finding closure to their fertility journey (Franklin, 1997; Becker, 2000). This imperative is intensified with the proliferation of optional choices. My previous work (Perrotta and Hamper, 2021) emphasizes how patients have to navigate an increasing number of available options in their quest to find *the treatment or combination of treatments that will work for them*, as one of the patients interviewed elaborates:

> 'You're paying money for something which cannot be guaranteed. And it's a lot of money and you know, and it's … And it works a third of the time. So how, you know, in any other industry, like how on earth could that add up. Like why would people do it? You know, it's, it's crazy but we do because, you know … Why are people so invested in it and therefore why do they want these add-ons, you know. Because a lot of it is you're just paying a lot of money for hope and it, you know, and the add-on is that, isn't it?'

The excerpt underscores the essence of operating in what I term a *hope market* as a consumer. In light of low success rates, to keep alive the hope of having a biological child, patients have to accept the high cost of treatment without any guarantee of success. Add-ons offer additional options to be explored, or at least considered, in the attempt to try anything possible, adding to the burden of treatment decision-making in a field characterized by high level of uncertainty.

As the interviewee emphasizes, due to the limited public funding and the high cost of treatment, financial considerations are central in patients' decision-making. For patients who undergo NHS-funded treatment, even when these additional interventions are used, they do not represent a cost. However, undergoing these interventions in NHS-funded cycles legitimizes their use beyond the NHS and creates expectations as well as the conditions to request the same treatment as a private patient, even if it entails an extra expense (Perrotta and Hamper, 2021).

For patients who self-fund their treatment, the cost of these interventions is weighed against the cost of the whole procedure. For instance, the same patient discussing her decision to include one of these interventions comments:

'I think sometimes there's this bit of a thing oh, add-ons, you know, that everyone, that people who are having it are just sort of daft and have no idea what they're doing. Whereas I think that people have every idea what they're doing but they just want to have a baby and if you say to them well, if you spend £500 on this and scratch, whatever, you … Well, it might, it might help … Then you probably pay it, wouldn't you? It's like in the grand scheme of things it's not that much. But I think that's quite hard for people to understand and then they worry that people are being exploited and things like that. But I think the whole industry is, in its own way for what it is, exploitative so it's sort of like why, why draw the line anywhere in it, you know?'

As this patient emphasizes in the interview, the additional cost that these interventions represent is quite marginal compared to the cost of the whole treatment. As she points out later in the interview, when "you're already paying ten grand, it doesn't matter" if you add a few hundred pounds. It is crucial to underline that, from this patient's perspective, deeming add-ons as the focal point of exploitation appears entirely arbitrary, given the exploitative nature of the entire experience. In a similar vein, she underlines that initiating fertility care requires a certain level of trust that the entire medical procedure is safe and effective:

> 'I'm already paying all this money to inject myself with loads of drugs that may not work so you know, you will just have to hope that it's being regulated enough that you're not, you know, going to get completely screwed over.'

As the view of this patient underlines, the cost is not the only factor to consider and several other elements have to be put into perspective, including a certain acceptance of the experimental nature of the treatment, its low success rate, and the need for multiple attempts (Perrotta and Hamper, 2021). It is crucial to note that, although the pursuit of parenthood now offers a wider array of options, this does not necessarily translate into an increased likelihood of becoming parents through fertility treatment. Indeed, while the success rate of fertility treatments has increased steadily over the years, it remains low overall, with a 22 per cent live birth rate (LBR) in the UK (HFEA, 2023).

The uncertainty and ambiguity that characterizes the experience of infertility (Sandelowski, 1987) remain unresolved by fertility treatment. As others have noted (Allan, 2007), paradoxically while fertility treatment enables patients to endure ambiguity and uncertainty by offering hope, it simultaneously contributes to the emergence of additional ambiguity and uncertainty. For some patients, closely directing their treatment serves to attempt to gain a sense of control, even if illusory. For instance, a patient who had a baby from her first cycle ponders whether she would have paid for one of the interventions she was offered if her treatment had not been successful:

> 'Yeah, we would have found that money somehow because I was also, we were, I was also paying for acupuncture, you know, we were prepared to throw some money at it for things that had ... In a situation where you

have no control, fertility, you have no control and … One of the overwhelming emotions is the lack of control and I knew it was beyond my control, so I knew that I had to take any steps that could be within my control I had to take them. So that's one of the things that I was able to take control over, right, well, we can make a decision that will possibly help improve the chances so we'll take it.'

It is important to emphasize that not all of those who are offered add-ons will want to use them. For instance, a patient going through treatment with an unexplained subfertility diagnosis elaborates on how, despite the appeal of these interventions to offer answers, they decided not to opt for any additional interventions for financial reasons:

'I was aware that, you know, these are, some of these are just companies saying oh, we can do this and we can do that and it's, you know, it's money-making some of it, isn't it? And a lot of them do prey on people who are desperate. But it, at the same point my medical knowledge is not that of a doctor so I can't, I don't know. But when you're being told there's a chance it could be this, you think, oh well, why can't I be tested for that? And at the same point I didn't have a couple of thousand pounds spare to take out of potential needing IVF money to go and spend on all of these tests.'

While financial concerns were common, some patients present varied justifications for opting against the use of these interventions after careful consideration of some of those they were offered. For instance, following three unsuccessful cycles, a patient shares their experience of exploring a visit to a private clinic:

'They [private clinic] were trying to offer us lots of add-ons that we knew from the research we'd done beforehand were proven not to … And that NICE were recommending that shouldn't be offered but they were kind of, there was a little bit of that.'

In this patient's case, feeling pressured into purchasing unproven interventions prompted them to opt for self-funded treatment at the same NHS clinic where they were previously treated. While a minority of patients mentioned feeling influenced, the majority of those undergoing treatment in private clinics emphasized the lack of such pressure. For instance, a patient undergoing treatment in a private clinic states:

'I think the one thing I can say is that [our clinic] never encouraged add-ons. I said "Do you think I should get the scratch?" And they said "That's up to you." They go "There is no clear evidence to say that it works but there's no clear evidence to say that it doesn't." So that was my choice from what I have read to get that done or what other people had maybe shown me. So from a clinic point of view no … I don't know. It's hard to explain. I think it was just that I'm, it was my choice. No one pressurized me but what I was reading there was totally mixed things and who was to say that it helped at all, you can't, you can't know or not know if that is the reason why we got pregnant or not.'

This excerpt illustrates how patients' expectations regarding these interventions are not formed in isolation, but rather shaped by how these are presented through various sources. As highlighted earlier, it is vital to emphasize that patients have diverse needs, priorities, and sensitivities. As a result, not all patients presented with the choice decide to pursue or invest in these interventions. In the next section, I will delve into the perspectives of those who opted to pay for unproven treatments, despite being conscious of the lack of evidence supporting them.

Buying potential

As anticipated, a common assumption in the medical and public debate is that patients are either uninformed about the absence of evidence or do not comprehend what should be considered robust evidence. However, as I have shown in the previous chapter and has been confirmed by other studies, many patients who are painfully aware of the lack of evidence have expressed a desire to opt for these treatments, even when they come with a high cost. As this patient elaborates:

> 'I suppose something like that [an optional intervention] where if it's going to, if it's going to improve, if it's clearly proven that it's going to improve the chances of someone getting pregnant, then maybe it should be standard rather than an add-on. But if it's not proven and there's a potential then that person should have the choice.'

As the excerpt emphasizes, even though a treatment is not proven yet, in the logic of hope this does not exclude its *potential* to work. The intimate link between hope logics and potentiality has been explored in the social scientific literature on fertility and broader biomedical innovations (Franklin, 1997; Taussig et al, 2013). The concept of potentiality creates space to accommodate uncertainty and unknowns, therefore fostering the promises of hope for a technologically and biomedically enabled future. The emphasis on the *potentiality* of biomedical innovation is very common in the narratives of patients who decide to purchase these treatments. For instance, while discussing their decision to opt for an unproven treatment, this patient explains:

> 'That [EmbryoGlue] was another option that was offered to us. And again, I'm quite a bit of a researcher so I go onto the HFEA website and they are on orange, red or green to see, you know, what the professional body says

about it. But again there's no really clear evidence that it can help. But you just invest anything you can, because there is that slight thought in the back of your mind that it might help. So yeah, that's what we've done as well on both the cycles ... I don't know, it's really hard to explain, it's why would you not do it? It's like saying this one might give you a baby.'

The overwhelming power of attraction of these treatments lies exactly in the intangible but undeniable and undoubted potential they hold that it *might help*. The trope of potentiality directly mirrors, and some studies suggest is shaped by (Carson et al, 2021), one of the dominant professional discourses explored in the previous chapter. This discourse revolves around professionals questioning whether the lack of evidence equates to evidence of absence. Likewise, patients wonder whether the fact that a treatment is not proven actually means that it does not work. While many patients are painfully aware of the lack of evidence, the appeal of the potential remains untouched. As others have noted, in the context of fertility care 'doing everything possible' is problematic, 'because the technology remains one of seemingly endless possibility, always offering the "maybe next time" promise of success that postpones the end of treatment' (Throsby, 2010, p 239). The growing number of novel interventions that can constitute treatment carry additional layers of potentiality that particularly appeal to some groups of patients. For instance, one of the patients interviewed recounts how her view regarding treatment, which conflicted with her doctor's view, was guided by her goal to have more children through fertility treatment, not just one:

'So his first reaction was that he really doesn't think I should be focusing on having more children, that the goal, I'll be lucky if I have one. And I was really insistent, explained to him that it is important for my mental health as well. We don't care if you know, for any evidence or medical reasons, I need to do it for myself. I need to know I've done all I could.'

Similarly, in discussing the physical pain associated with endometrial scratching (ES), a patient stressed that "if these three seconds of pain help me have a baby earlier than expected then yes, I'll do it". While the previous patient was planning ahead for her future family, this patient's considerations regarded the potential ability to reduce the time needed to have a baby.

Although individual patients' priorities vary, a common perspective was that the appeal of the optional interventions increases through the treatment path. For instance, discussing the ES, a patient at their first attempt remarks:

'I think I would try it if the IVF failed. I think that's kind of like a natural progression, isn't it? Then you start searching for other things that might just work. But at the moment because I've not done one, I think I will just do a straightforward transfer. See what happens.'

This "natural progression" is part of the hope narrative described earlier. As Franklin (1997) highlights, paradoxically, every unsuccessful treatment cycle is seen as offering fresh clinical insights that will enhance the prospects of success in subsequent cycles, rather than an indication that the treatment is ineffective. Indeed, the appeal of the potential gets stronger when patients perceive optional treatments as their last resource, either after many unsuccessful cycles or for older patients approaching reproductive decline. To emphasize the extent to which this last resource is perceived, one of the interviewed patients speculates:

'I think if you have a cancer patient who's gone through all the treatment, chemo thing and they've been given six months to live and nothing is working and then someone says there's a doctor in Holland who's doing some research and they think this might help, you know, it's going to cost you £20,000 to try it, I'm sure they … And I think desperation is the right word. It's true because it's, I think at some point, well, you … when you weigh up the pros and cons the biggest pro is that it might work, you know, there's that 0.01 per cent chance that it might work so why not take that risk? And I think that's the same [in fertility treatment] … If you don't do this treatment you probably will never have a baby.'

Although the comparison with cancer might sound rather extreme, the excerpt underlines an aspect that characterizes fertility treatment. Fertility patients do not risk death from unsuccessful treatment, but that treatment might represent for them their last chance of having a biologically related child. To stop trying implies surrendering not only the hope of becoming a parent but also the fulfilment of a set of life expectations and experiencing significant disruptions to life plans (Becker, 1999). Infertility can be a source of profound grief, as it deprives someone of the future they had imagined. Individuals who experience involuntary childlessness must contend with the trauma of a disruption to their life narrative and a lack of direction (Becker, 1999; Kirkman, 2002, 2003).[7] Undergoing fertility treatment helps to progress toward the next step of the infertility path, that might comprise a child or not, but represents a form of hopeful expectations of a better and resolved future (Franklin, 2022). This circumstance renders decisions to forgo treatment highly challenging (Carson et al, 2021), and, similarly, makes it difficult to dismiss the idea of considering optional treatments.

The perception of optional treatments as the last chance might not be due to the number of previous unsuccessful cycles of treatment, as the urgency increases with age. As discussed earlier, the average success rate rapidly declines when individuals are in their thirties. Therefore, as the same patient notes:

'I think it depends on the age, like I said, if a couple is in their early twenties, they might not be so desperate to do it, if a couple is in their late thirties/early forties they probably would be a bit more desperate.'

The cycle of hope and despair is a recurring motif in the quest for happiness through fertility treatment (Morris, 2019). This cycle alternates between moments of promise and periods of pain, which require further hope in the potential of the biomedical intervention. As Sarah Franklin (1997) highlights, the trope of 'desperateness' has been long used both in the media representation of and public discourse on IVF to legitimize the procedure in the UK. As she points out, if fertility treatment has the potential to assist individuals in attaining their desired child, why would anyone want to deprive them of this hope or hinder their experience of joy? Similarly, some interviewed patients used the desperate desire for a baby to explain the need to have hope in the potential offered by these interventions.

It is crucial to notice that many of the interviewed patients who opted and paid for optional unproven interventions were aware of the lack of evidence and had often carefully engaged in an individualized assessment and criticisms of evidence themselves.[8] For instance, a patient summarizes the challenges of evidence production discussed in the previous chapter:

'If everything has to be evidence based then we exclude a lot of treatments or opportunities that are very difficult, can be difficult to prove with the gold standard study. There could be a lot of smaller studies done that show that this works. Or there could be many singular, you know, personal successes that show that this works, but whether it's for practical reasons or financial reasons, funding reasons, they may not have that gold standard evidence study available.'

When asked about what this gold standard evidence study would entail, the patient further explains that "They have to do a particular type of research study with specific number of participants etc. and fulfil, there are criteria and it's not that easy to conduct a study like that". While, as previously discussed, this narrative on the challenges of evidence production closely mirrors the professional discourse, patients often use it to legitimize the individual choice of treatment despite the lack of evidence.

For many patients, the combination between the challenges of evidence production and the motif of 'trying anything possible' makes these interventions and their potential a perfectly rational choice to keep going forward. As conclusive evidence is not available, other forms of evidence have to be taken into consideration and this sometimes includes anecdotal evidence. For instance, discussing the case of reproductive immunology (rated red on the HFEA rating) a patient emphasizes:

'It's just so discredited in the professional community and almost some doctors who offer it are seen as charlatans, but it worked for so many women that I talked to, who did rounds and rounds of IVF and nothing worked, unexplained infertility, and a lot of them had high thyroid antibodies like me. Made me think it could be a link and yeah.'

While references to anecdotal evidence were relatively common among interviewed patients, their weight in the decision regarding treatment was, in most cases, marginal. However, these stories further fuel the uncertainty and support the potentiality of these interventions, emphasizing that they might work. These circumstances, as a patient notes, can produce what she calls *the next one might work kind of trap*:

'I think it just kind of made me a bit more like aware of how "the next one might work" kind of trap, do you know what I mean? Like spending more and more on it and yeah, like there's no guarantees with any of those things, but you can very, very easily be swept into paying for things, extra things when you're desperate to have a baby as well.'

While patients acknowledge the possible financial consequences, some find it challenging to establish clear boundaries regarding which interventions to contemplate, select, and fund. The focus on potentiality and hope in the face of failure and uncertainty raises the risk of overlooking the financial, physical, and emotional costs of undergoing fertility treatment. As I have investigated elsewhere (Perrotta and Hamper, 2021), for individuals who are not successful in achieving their desired outcome, hope becomes a complex challenge that has to be managed. For many patients, their primary considerations revolve around potential harm. For instance, when explaining her preference for a particular clinic, one of the interviewees observes:

'We are phobic of those big private clinics. I will not touch them with a barge pole, so I'm quite different to a lot of other people but ... I'm very vocal about how I feel about that on my blog because I just think there's a lot of unsafe medicine being practised and I don't like it. And I like where we go because he [the doctor] is so transparent, he won't let me do anything that could potentially cause harm. He will let us do things that don't have a strong evidence base, but as long as then nothing is potentially going to cause harm, like for example we paid to have the ERA testing done ... But he won't let me have steroids or this or that ... We know that there is no evidence [for EmbryoGlue] but we know that it doesn't cause harm, you know. And they know that people are working on studies and that there might be kind of breakthroughs so they just do it.'

Both the use of steroids and the endometrial receptivity array (ERA) test, a method claiming to find the optimal time for embryo transfer, are rated red on all iterations of the HFEA rating system. Interestingly, while this patient uses these examples to draw a distinction between potentially harmful and safe unproven interventions, the HFEA rating for ERA is red 'because the findings from moderate/high quality evidence shows that this add-on may reduce treatment effectiveness'.

While for some patients harm is where a line should be drawn, for others embracing a certain level of potential risk is an integral aspect of the fertility treatment process. For instance, while discussing the forms she was sent by her clinic regarding the potential risks of the whole treatment, a patient emphasizes:

'I didn't really read a great deal into things. I just sort of, you just sort of do it because you think, oh, this is giving me an extra chance. So you obviously read the dangers and things and think, well, what are the dangers of IVF and it does, you do scare yourself thinking certain cancers and things like that … I did mention it to my doctor because just before I had the IVF I felt like a little lump in my breast and it was a cyst … And then I thought, I wanted a baby, so you sort of take the risk sort of, I think. And the fact that we didn't have any other way of getting pregnant, I think we just thought, well, whatever risks are there, we'll just have to deal with it.'

Although this patient was concerned about the potential risk of cancer because of her previous experience and discussed this risk with her doctor, eventually she decided to undergo treatment anyway because she felt that treatment was the only way to achieve her desired baby.

The examples presented in this section emphasize the relation between the promissory narratives that characterize scientific progress and biomedical innovation in this and other medical fields, and patients' willingness to buy the additional potential offered by optional interventions. However, they also raise emerging concerns about an uncritical use of the concept of informed choice in making treatment decisions. I will delve deeper into these concerns in the next section.

Informed choice in the hope market

As discussed earlier, patients have different approaches to treatment decision-making: some prefer not to take responsibility for their treatment and opt to follow their doctors' recommendations, while others wish to be actively involved and, at times, even direct treatment decision-making themselves. Beside personal preferences, most patients underscore their right to informed choice, especially when treatment is self-funded. According to many patients, this includes the right to choose the optional interventions they pay for, as the decision to explore the potential, as investigated in the previous section, should ultimately rest with those undergoing treatment.

Similarly, as discussed in Chapter One, many professionals agreed with this view and supported the right of patients to choose unproven treatments, provided that the absence of conclusive evidence was clearly explained to them. However, conflicting viewpoints emerged regarding the extent to which patient choice should be promoted, given the implications encompassing both medical and financial dimensions. For instance, while the majority of professionals endorsed the concept of patient choice, the scope of what constituted a choice varied based on the context, the interpretation of medical practice, and individual patient circumstances. Alongside a robust pro-choice narrative, numerous treatment decisions were, in practice, made by professionals and presented as standard protocols or medical recommendations. A notable example is one patient who, after careful consideration and evaluation of available evidence, decided against choosing any add-ons, and shared their perspective on the inclusion of intracytoplasmic sperm injection (ICSI) in their treatment:

> 'I guess it [ICSI] could be [an add-on]. I guess I'm not thinking about it as an add-on because that's what they recommended so it wasn't, it wasn't given to us as an option, it was given as "oh, we strongly recommend this" … Because we had a low, we had five eggs collected first time round but only had one fertilized so … And then with the ICSI we have had a lot more fertilized than the previous, in the last two rounds so. So yes, so I guess, in some ways I guess it is an add-on to the original treatment but it wasn't presented to us as a … It was presented as a recommendation rather than a …'

Within any clinical or medical practice, the array of choices that patients can opt for is significantly restricted, influenced not only by medical expertise but also by the governance model of healthcare organizations. In clinics operating under the NHS, while patient empowerment is acknowledged as a valuable principle, the tangible options extended to patients remain restricted. Consequently, choice is confined to a handful of carefully evaluated alternatives. On the other hand, some private clinics, seeking competitive differentiation and aiming to provide patients with broader choices, significantly expand the number of available options. In addition, in a sector characterized by high uncertainty due to a widespread lack of conclusive evidence and with low success rates, making decisions about treatment implies taking significant responsibility for the potential negative consequences associated with the high risk of failure. As I detailed throughout the previous chapters, these variations in what is considered optional introduce additional tension between NHS and private care.

In the context of this research, although informed choice was endorsed by most patients, the range of options presented as choices left some patients perplexed. For example, while discussing her treatment, a patient recounts:

'I had read something online about taking Metformin and I suggested that to the consultant and they said yes. So, I was like, is it a good idea? That is when I was like, "what on earth is going on here". I was like, "okay, I'm suggesting something here and they said yes, sounds like a good idea".'

The excerpt highlights the inherent tension between the patient's desire to influence her treatment and her astonishment at the doctor's sincere consideration of her medication suggestion. While discussing her interaction with the consultant at a potential new clinic, following two cycles at a different centre, another patient remarks:

'We'd had two IVFs, we'd had two fresh cycles, one was a long protocol and one was a short protocol. The long protocol worked obviously a lot better for me and I asked him [the consultant in a potential new clinic], "What kind of protocols do you use in your clinic?" He said "We only do the short protocol." And I said "Oh, is there any chance I could do the long protocol?", because obviously that worked better for me and he sort of brushed it off. And then the next time, we saw him he said something along the lines of "I recommend the short protocol, but if you want to do the long protocol because it worked for you in the past you can do it." And again, I sort of sat there with my husband thinking that I'm spending ten grand and you're asking me to pick what protocol I'm going to do.'

While patients typically do not have the option to choose their stimulation protocols, this case illustrates how an excessive emphasis on patient choice is not only professionally questionable but can, ironically, also prove to be counterproductive. In this instance, the patient interpreted the professional's attitude as a clear sign of vested financial interest. In fact, the interviewee continues:

'It felt like it was money to him. And you don't say to someone "well, I'll put you on that protocol because it makes you feel better", because it's not about that. It's about what works for your body ... And it's just stuff like "do you want the Cytakines [proteins produced by immune cells] to be tested? Do you want this? Do you want that?" I just think it's so wrong that they get to ask you when you're so vulnerable with all this money [involved], as these tests are not cheap. So, yeah, and I think I'm in a privileged position that I'm a health professional and I just think some people, you know, wouldn't be in that position.'

Given her background as a health professional, this patient is more openly critical of the medical practitioner. However, the perception of being offered choices that are considered inappropriate was shared by several patients.

In a context marked by uncertainty, delegating treatment decision-making to patients carries the risk of eroding trust in medical professionals due to concerns about their vested interests in profiting from treatments.

While some patients do not want to be burdened with this additional responsibility and expect to be able to trust their doctors, others approach treatment with a consumer mindset. For instance, one of the patients interviewed shared her experience of approaching a clinic with specific expectations regarding the interventions her treatment should have included, such as immunology treatments (rated red by the HFEA).

However, the clinic opposed this patient's demands and recommended to start with a standard treatment:

'In that first appointment she [the consultant] did answer some of the questions and then we started discussing this autoimmune thing. I was really keen that they gave me treatment to suppress my immune system, which they really did not want to do. They said that we needed to have a good try at IVF, so the first round if it didn't work then we'd see where that was. In the meantime, I'm thinking that's going to cost us £10,000 and for you to just say no, you can't have this treatment that might work … But we put, we did put it in their hands. We went with what they advised and I do feel as well like that was another thing that made us decide that we didn't need to look at another clinic, because they could have upsold us loads of things and they didn't. And they said this is not necessary, that is not necessary to the things that I was saying, okay, I've read this is good, I've read that's good. And it was, all of it they were just like look, you just need a normal round of IVF, see how that goes.'

As this patient points out, their inclination to 'try everything' was met with resistance from the professionals overseeing their treatment. Initially, the patient was outraged by this refusal, particularly given the high cost of the treatment. However, allowing professionals to take responsibility for the treatment ultimately helped restore a sense of trust in them. Upon further reflection, the patient interpreted this approach as a genuine medical recommendation and a confirmation that the professionals were not driven by financial interests.

The examples presented in this section reveal the highly problematic nature of informed choice in this context. Patients are not only held responsible for selecting certain interventions but also for covering the associated costs, despite the lack of substantive information to inform these decisions. While the provision of accurate and transparent information remains paramount for enhancing patients' experiences, as I will discuss in the next chapter, it is imperative to explore additional regulatory strategies to address the commercial aspects of fertility care.

FOUR

Regulating the Hope Market

The add-on rating system explored in Chapter One was one of the regulatory measures implemented by the Human Fertilisation and Embryology Authority (HFEA) as a response to the heated debate on add-ons. Before delving deeper into the regulatory authority of the HFEA within the field of fertility care, it is crucial to stress that these measures coexist within larger regulatory frameworks overseeing medical interventions. These frameworks are intricate and exhibit notable variation across different countries.[1] Broadly speaking, distinct procedures and approval criteria govern medicines and medical devices. In the UK, both fall under the regulation of the Medicines and Healthcare products Regulatory Agency (MHRA).

Medicines[2] must obtain marketing authorization, also known as a licence,[3] before being marketed, and they require a prior clinical trial authorization before submitting a licence application. To obtain a licence, which outlines the approved supply conditions and patient groups for it, the medicinal product must meet the relevant standards of safety, quality, and efficacy.[4] The effectiveness of a medicine must be demonstrated to surpass that of a placebo. However, there is no requirement to prove its superiority over another product licensed for the same purpose, let alone a recognized, first-line treatment (BMJ Publishing Group, 2009). Moreover, the authorization does not restrict the medicine's use to specific purposes, allowing de facto healthcare practitioners to prescribe it for off-label use.

The MHRA also regulates the UK medical devices market, which includes in vitro diagnostic medical devices (IVDs).[5] Prior to being made available in the UK, medical devices must possess either a UK Conformity Assessment (UKCA) mark or a European Conformity (CE – Conformité Européenne) certificate.[6]. This certification[7] serves as evidence that the device complies with relevant regulations. Manufacturer companies are responsible for ensuring that medical devices maintain the required standards of safety and performance after they have been introduced to the UK market. Additionally, manufacturers are required to submit vigilance reports to the MHRA whenever certain incidents involving their devices occur in the UK.

Within the field of fertility care, this means that the HFEA's regulatory scope is limited in regulating medical treatments. For instance, the MHRA is responsible for regulating the composition and safety of culture media used in British fertility clinics. When research in this area raised concerns about culture media composition and its potential impact on embryonic development, the HFEA Scientific and Clinical Advances Advisory Committee (SCAAC) monitored these findings to inform the MHRA.[8] However, neither the SCAAC nor the HFEA have the authority to intervene directly in the regulation of culture media. The involvement of multiple regulatory bodies, each with specific responsibilities and limitations, makes the regulation of biomedical innovation within the fertility sector intricate and multifaceted. I will delve into this further in the next section.

Regulating medical innovation in the fertility market

The UK fertility sector is often represented as a carefully regulated but permissive environment (Roberts and Throsby, 2008), where Parliament establishes the rules and the HFEA has the delegated power of quality control and evaluating scientific progress (Callus, 2011). The HFEA was the first authority of its kind and since its introduction, established by the Human Fertilisation and Embryology Act 1990, has represented the gold standard to regulate fertility care (Cutting, 2012).

The HFEA's role encompasses the examination of both public and societal concerns, as well as the interests of all the stakeholders it oversees, including fertility clinics, professionals, and patients. Navigating within such a vast legislative framework inevitably results in tensions arising among various groups' interests. In this context, the HFEA must carefully balance the imperative of protecting patients and society with the goal of promoting responsible research and medical practices (Morgan, 2004).

The tension between often contrasting interests has generated heated debates in several cases. Over the years, the authority has been criticized by scientists for its conservative approach towards innovation (Meikle, 2004; Savulescu, 2011), contested by some patients for refusing approval for particular treatment applications, and accused by some lobby groups of going beyond its statutory powers (see Callus, 2011).

The constant technical and scientific development in a field characterized by ethical dilemmas and competing interests has required a review of the Act, which came into force in 2008 (for details, see Cutting, 2012). Similarly, as of the time of writing, the authority has embarked on an additional review of the law to modernize the regulation of fertility treatment and research.[9] The controversy around fertility treatment add-ons is one of the pillars in the HFEA's case for change. The open consultation on modernizing fertility law launched in 2023[10] proposes a request for a wider and more balanced spectrum of regulatory powers, including the establishment of mechanisms to promote innovation through the authorization of trials and the verification of initial claims. During her speech introducing the proposed changes to the law, the Chair of the HFEA (Chain, 2021) highlighted the necessity for implementing new forms of regulation in an increasingly commercialized sector. These new forms of regulations should aim to ensure the protection of patients in their capacity as customers and include influencing clinics' financial interests by imposing economic fines. While the HFEA's intention is not to prohibit treatment add-ons or impose fines on clinics that provide them, the focal point of these proposals is addressing the poor quality of information on add-ons and the risk of mis-selling them.

Concerns about the quality of information provided to patients had been raised well before the emergence of the add-on controversy. As an illustration, in their 2015 online guide for patients titled 'Getting Started: Your Guide to Fertility Treatment', the HFEA recommended posing questions that evaluate the evidence supporting these interventions. Some of the questions patients were encouraged to ask their clinics included, 'Is this treatment recommended by the NICE and, if not, why? Has this treatment been subject to randomized controlled clinical trials that show that it is effective and is there a Cochrane review available?'

The focus on patient empowerment and information transparency was further strengthened through subsequent regulatory measures, such as the implementation of the add-on rating system in 2017 and the development of the add-on consensus statements in 2019 and 2023, as discussed in Chapter Three. These regulatory approaches were the authority's primary immediate means, given its constrained ability to halt clinics from offering these interventions or regulate their market. Before delving into the discussion of these issues in the following sections, it is crucial to emphasize that the HFEA is the competent authority to authorize most of the processes falling under various licensable activities in the Act, but its power to directly regulate specific interventions is restricted. Under the current legal framework, the HFEA engages in licensing clinics, conducting regular inspections, and offering guidance on how to meet legal obligations through their 'Code of Practice'.[11] While the authority has put in place a system for licensing novel laboratory processes for use in fertility treatment,[12] and regularly reviews them, it does not possess the authority to authorize or regulate the use of medical devices and medicines themselves, which are overseen by the MHRA, as previously discussed.

Table 4.1 provides a comprehensive list of licensed activities and authorized processes at the time of writing.

Table 4.1: Licensed activities and processes authorized for use in clinical practice

Licensed activity	Authorized processes
Procuring gametes	Egg/ovarian tissue collection Surgical sperm collection
Keeping gametes	Culture of eggs
Processing gametes	Semen preparation (including the use of reagents to increase sperm motility) Egg preparation In vitro maturation Thawing/re-warming gametes Egg activation using calcium ionophore (only in suitable patients[a])
Distribution of gametes	Transfer of egg/sperm between centres
Use of gametes	IUI/GIFT/IVF/ICSI
Storage of gametes	Freezing of eggs/sperm Vitrification of eggs Freezing of testicular tissue (not for transplantation purposes unless a HTA licence is in place) Freezing of ovarian tissue (not for transplantation purposes unless a HTA licence is in place)
Storage of embryos	Freezing of pronucleate embryos/early cleavage embryos/blastocysts Vitrification of embryos/blastocysts
Creation of embryos	IVF/ICSI
Procuring embryos	Lavage

(continued)

Table 4.1: Licensed activities and processes authorized for use in clinical practice (continued)

Licensed activity	Authorized processes
Keeping embryos	Culture system
Testing embryos	PGT-M/PGT-A Polar body biopsy
Processing embryos	Culture Assisted hatching (mechanical, chemical, laser) Morphological grading Manipulation Thawing/re-warming of blastocysts and embryos Non-invasive assessments Intrauterine culture of gametes and embryos (including insertion and removal of device, followed by transfer of embryo[s] to the same woman)
Distribution of embryos	Transfer of embryos between centres
Placing permitted embryo in a woman	Embryo transfer
Using embryos in training	Embryo/blastocyst biopsy Cryopreservation and thawing techniques Vitrification Assisted hatching (mechanical, chemical, laser) Embryo handling and manipulation Assessment of embryos

Note: [a] A note to this treatment available on the same website specifies that 'The HFEA's Scientific and Clinical Advances Advisory Committee considered the use of Calcium Ionophore as an egg activation technique and highlighted the theoretical risks relating to embryo viability (e.g. premature activation and triploid embryos). Given the theoretical risks of using Calcium Ionophore, centres using it are expected to do so only in selected patients, such as those with PLCz deficiency. Centres are expected to document their rationale for using Calcium Ionophore for individual cases. As with all treatments and processes, centres should ensure that patients are fully informed about the efficacy and potential risks and that validation is carried out.'

Source: Adapted from HFEA 'Authorized Processes' table, available at https://portal.hfea.gov.uk/knowledge-base/other-guidance/authorised-processes/ (last accessed 30 July 2023).

The SCAAC and the Statutory Approvals Committee (SAC) grant these authorizations and regularly review them based on new evidence. Additionally, the Code of Practice goes into detailed discussions about the lawful execution of certain licensed activities, such as intracytoplasmic sperm injection (ICSI) and pre-implantation genetic testing for aneuploidy (PGT-A), and outlines criteria for selecting and using laboratory equipment and materials (see guidance note 26 v.2). While the Code of Practice stipulates that only medical devices certified in accordance with UK or EU standards should be used, the responsibility for defining these standards does not lie with the HFEA. Instead, the standards for medical devices are typically set by relevant regulatory bodies or authorities responsible for medical device certification. Similarly, the Code references a series of HFEA clinic focus articles to define the use of specific medical treatments, like the off-label use of intralipid infusions.[13]

Given the existing constraints, the current suggestions for legislative amendments include a proposal to introduce relatively low-risk novel interventions subject to closer, real-time regulatory monitoring. This approach aims to broaden the scope of the HFEA in regulating the introduction of biomedical innovations, which would facilitate the simultaneous accumulation of additional evidence to support treatment efficacy and improve patient safety.

Fairness of information to make informed decisions

As mentioned earlier, concerns have been raised about the quality and reliability of information on add-ons, and fertility care more broadly, on a global scale. With the diffusion of the internet, the influx of information available online from various sources has proliferated (Marriott et al, 2008). Particular attention has been paid to fertility clinic websites, as these are considered reliable sources by patients. For instance, an early study (Abusief et al, 2007) assessing US fertility clinic websites' compliance with joint guidelines for advertising established by the American Society for Reproductive Medicine (ASRM) and the Society for Assisted Reproductive Technology (SART) in 2004 showed very limited compliance by both private and academic clinics. More recent compliance reviews of US clinics (Chan et al, 2014; Sauerbrun-Cutler et al, 2021) show that over 15 years later the situation has only marginally improved. As an example, many websites fail to adhere to core guidelines surrounding reporting in vitro fertilization (IVF) clinic success rates.

A growing literature examining fertility clinic websites globally (Kadi and Wiesing, 2015; Wilkinson et al, 2017; Goodman et al, 2020; Carneiro et al, 2023) also shows that the transparency and quality of information provided is highly heterogeneous and often not very accurate. In particular, data on success rates is often reported in potentially misleading ways. Unsurprisingly, more recent studies that have focused on the provision of information concerning add-on treatments have shown similarly poor results (Heneghan et al, 2016; van de Wiel et al, 2020; Galiano et al, 2021; Lensen et al, 2021b; Perrotta et al, 2024).

Many of the patients interviewed in this research discussed the absence of clear and accurate information about these treatments. They expressed concerns about the misleading and confusing effects of the vast amount of available information, which lacks clear guidance on its accuracy. For instance, discussing the HFEA rating a patient comments:

'The HFEA website definitely, I think it's going down the right route. I think it needs, you know, it would be great if it had more information. You just want, yeah, you want a body of people to just give you really clear guidance and it's just … Because it's such an emerging, constantly changing and emerging industry that that's, that's something that is quite hard to keep on top of, isn't it?'

As the interviewee highlights, the ongoing evidence production and availability of new information makes it difficult for patients to be up to date. The appetite to seek and assess information, as well as evidence as I discussed earlier, varied significantly across the group of patients who participated in this research. Patients who were more inclined to seek additional information beyond institutional sources, such as the HFEA rating system, expressed significant frustration in evaluating the reliability of various sources. As this patient elaborates:

'There is no clear place to find any information. You just go down a rabbit hole of Googling and internet and American research and research from all around the world and most of it isn't, you know, peer reviewed. You know, it's just, it … a lot of anecdotal stuff and so it felt really, really stressful because it just sort of took over my brain for a long time, you know, searching, searching.'

The conflicting nature of the information available was discussed by many participants in this research. For instance, one of the interviewed patients states:

'One thing I sometimes find difficult is, you know, obviously you look and I think it's the amount of information now that's available and you look online just for a bit of advice … Conflicting information is hard sometimes because you just think, right, okay, I've got my head around that and that and then you read something else and it's like oh! Back to square one again.'

Recognizing the need to increase the quality of information provided to patients concerning the potential benefits and risks of add-on treatments, the Competition and Markets Authority (CMA) took action in 2021 by releasing guidance specifically designed for fertility clinics (CMA, 2021a), along with a corresponding consumer right guide for patients

(CMA, 2021b). Moreover, the Advertising Standards Authority (ASA) has issued an enforcement notice (ASA, 2021) outlining guidelines for advertising IVF fertility treatment in the UK.

These guidelines specifically include information clinics should provide on their websites, as both the HFEA patient survey and research commissioned by the CMA showed that patients rely on these as trusted sources of information (HFEA, 2019; CMA, 2020, 2022b). These guidelines aim to ensure that patients are making well-informed decisions regarding the purchase of add-on treatments. As the CMA emphasizes, when patients pay for them, the commercial side of medical treatment has to abide by consumer law. Therefore, to protect customer rights, the CMA guidelines mandate fertility clinics to provide information on their websites about the risks associated with treatment add-ons, the available evidence base, and the HFEA's information concerning them. Additionally, clinics are required to direct patients to the HFEA's website for further information.

A year after introducing their guidelines for clinics, the CMA conducted a review of information on add-ons that were accessible on clinics' websites (CMA, 2022a). This review reveals instances of non-compliance with regards to the information offered concerning the examined add-ons (endometrial receptivity array [ERA], endometrial scratching [ES], EndomeTrio, reproductive immunology, and assisted hatching). The concerns primarily revolved around inadequate information about associated risks, insufficient clinical evidence, and misrepresentation of the HFEA traffic light system.

A systematic review of the information provided on time-lapse imaging (TLI) on clinic websites which I conducted with my colleagues (Perrotta et al, 2024) in June 2022 has shown similar results. For instance, over a third (35.8 per cent) of the websites stating that a clinic offered TLI did not provide any information on its cost for patients. In addition, almost half (47.9 per cent) of the websites did not mention or provide any links to the HFEA traffic light system. More concerningly, referring to early and mostly unspecified studies, a large proportion (42.2 per cent) of these websites made claims regarding the effectiveness of TLI that contradicted both the previous (amber) and current (black) assessment of the HFEA. While TLI is one of the interventions that is considered safer and non-invasive, these results show the serious lack of reliability and accuracy of the information available on fertility clinic websites. As TLI is offered by a third (67 per cent) of UK clinics, the consequences are far-reaching. Inaccurate and potentially misleading information not only affects patients' decisions to consider certain treatments but also poses a harm to both prospective and current patients who rely on clinic websites for trustworthy information.

Despite the limited level of compliance observed in these reviews, the regulatory efforts aimed at enhancing transparency and improving the quality of information have yielded positive effects in reducing the potential for mis-selling add-on treatments. For instance, a patient who was offered and declined various interventions explains:

'[I] read an awful lot of articles about what private clinics do try and push on you. And particularly going back to the NICE website but also the HFEA and seeing what do they … what studies are out there and if they're randomized. So we decided not to go for anything.'

While these regulatory approaches are indubitably a great step towards a more transparent and accurate quality of information, they have some limitations. As illustrated in Chapter One, a large proportion of patients are not aware of these resources and these groups are difficult to reach. In addition, a better quality of information will not improve the experience of those patients who want to delegate decision to their doctors. Finally, these interventions cannot solve the underpinning uncertainty generated by the lack of conclusive evidence and therefore do not fully cover the needs of those patients who want to be informed to make decisions.

Pricing strategies and treatment affordability

As discussed earlier, both the HFEA and the CMA have expressed concerns about the sale of expensive, unproven treatments to patients. However, their regulatory interventions have focused on ensuring the accuracy and transparency of information made available to patients, as the pricing strategies of clinics lie beyond their regulatory powers.

Similar concerns about billing practices were shared by many fertility professionals interviewed in this research. As explored in Chapter Two, many consider it acceptable for private companies to pass on the additional costs associated with using add-ons, if patients are well informed about evidence gaps. However, several professionals considered ethically problematic to overcharge patients to generate income from interventions not supported by robust evidence. For instance, discussing their experience in a previous clinic a medical director emphasizes:

'When [the clinic] introduced TLI and was charging £600–£800 for it without actually the evidence of it, you know … Why should patients pay for the clinical trials they were doing? I was very much against that.'

While most critics of professionals focused on charging patients for unproven treatments, others expressed reservations regarding some ethically concerning pricing strategies of their clinics, as these decisions are often beyond their control. For instance, some professionals stressed that some interventions such, as EmbryoGlue (EG), were overpriced to patients compared to how much they cost to the clinic. An embryologist we interviewed expresses this viewpoint:

'I think that I read something on the news not long ago about a particular clinic, can't even remember where it was, charging £500 or £600 extra just for using EmbryoGlue. EmbryoGlue comes in a 10 ml bottle. It is expensive, it's over £100 a bottle, but to charge patients for 5 ml like £500 or £600 is just ridiculous if there is no scientific evidence of something.'

As well as acknowledging that interventions like EG are more expensive than other available alternatives on the market, many of the professionals interviewed voiced concerns about how certain clinics, particularly in the private sector, engage in commercial practices that prioritize financial gain.

In a similar vein, a scientific representative from a company producing another TLI system challenges the principle behind charging patients a high fee (ranging from £300 to £850) per cycle treatment. As this interviewee suggests, clinics justify these high charges to patients by pointing out that TLI machines are expensive and that these costs need to be covered by patients. Indeed, a TLI incubator system can cost over £100,000, significantly more than standard incubators. However, while charging patients for a more expensive tool can seem justifiable, some pricing strategies remain controversial, as this representative explains:

'If it's a means to an end to cover the cost of the machine, fine, but … If you've paid off the machine but you're still charging them [patients] the extra £600, so then you're essentially just making a profit out of those patients.'

Some professionals also express concerns about the presence of ethically challenging practices associated with the pricing of additional treatments. As highlighted by a head of embryology, although their laboratory is equipped with TLI incubators only, an additional charge is still applied to patients who opt to use TLI. When asked about the consequences of patients opting not to undergo TLI, the interviewee explains:

'So what happens if they withdraw is that they're still culturing in that incubator but they don't get the video, they don't get the extra information, they don't get … And it puts the embryologist in an ethical dilemma … Well, you can't blind yourself from the information but then if you use it and they haven't paid for it and other patients have used it, what do you do then? And I think things which you know every patient should have because it's a standard of treatment, it shouldn't be an add-on.'

Similar views were expressed by representatives of the companies manufacturing some of the biomedical devices or laboratory equipment. For instance, a representative from one of the companies servicing TLI incubators highlights:

'If in a clinic you have the option of standard incubation and TLI incubation, then I guess it is an add-on if you choose to go with the TLI and the clinic is charging. Where I do have a problem is where clinics charge for blastocyst culture and for TLI because you're sort of almost, it's a double whammy. So that's where I think it's unfair.'

Blastocyst culture refers to the practice of extending the culture of embryos in the laboratory until they reach the blastocyst stage on day five of their development. While this has become a common practice in most laboratories,[14] some clinics still charge an additional fee of up to £800 per cycle (Heneghan et al, 2016) for extending the culture. In the case of clinics using TLI incubators, uninterrupted embryo development to the blastocyst stage can be achieved with relative ease. Therefore, as the interviewee underlines, charging patients for both the use of TLI and blastocyst culture as separate additional interventions is ethically questionable.

Interestingly, the discussion on the fairness of prices has been absent from public, medical, and social scientific debates on add-ons. While media narratives often focused on rapacious private clinics selling unproven, costly treatments to vulnerable patients (see, for instance, Schiller, 2019; Wilkinson, 2019), as discussed in the Introduction, the criticism was directed towards the lack of evidence rather than the overinflation of treatment prices. As I showed in Chapter Two, the emerging medical literature criticizing add-ons (Heneghan et al, 2016; Harper et al, 2017; Wilkinson, Malpas et al, 2019) similarly stresses the lack of evidence but overlooks the fact that some clinics use these interventions as a form of income generation. Moreover, given the absence of conclusive or generally high-quality evidence, this body of literature has never been compelled to address the logical consequence of its approach: What level of statistically proven yet marginal increase would justify the substantial pricing of these treatments? If one of these add-ons were to demonstrate a slight increase in success rates, would it then be justifiable to double the cost of the treatment for patients? As I showed in the last chapter, owing to the modest success rates of fertility treatment, the mere prospect of a marginal increase (illustrated with percentages as slight as 0.5 per cent above the existing success rate) could be enticing for some patients in certain circumstances.

As detailed in the previous chapter and confirmed by recent health economics research (Keller et al, 2023), one of the challenges of being a consumer in a hope market is that prices can be easily inflated, given their limited impact on demand. While this scenario explains the growing attraction of private equity firms to the fertility market, the different strands of literature on fertility have overlooked the consequences of unregulated pricing strategies in fertility care.

In contrast, an emerging theme of discussion in both the medical and social scientific literature is the so-called 'low-cost' IVF. For instance, in 2013 a European Society of Human Reproduction and Embryology (ESHRE) press release was issued stating that 'a study performed in Belgium has shown that low-cost IVF for developing and poor resource countries is feasible and effective, with delivery rates not much different from those achieved in conventional IVF programmes. This proof-of-principle study, say the investigators, suggests that infertility care may now be 'universally accessible' (ESHRE, 2013). This novel method essentially bypasses the need for a costly IVF laboratory by simplifying embryo culture methods, eliminating high-end equipment, and using inexpensive ovarian stimulation protocols. However, some procedures, like ICSI for instance, cannot be low cost as they require expensive laboratory equipment and highly qualified personnel to be performed (Ombelet and Dhont, 2016).

The Belgian team reported the first seven babies born through this method in 2014 (Van Blerkom et al, 2014) and has since launched 'The Walking Egg Project', an initiative aimed at increasing access to affordable treatment to a wider range of people around the world (Dhont, 2011; Ombelet, 2013, 2014; Ombelet and Goossens, 2016). While the notion of affordable fertility care has gained some traction in the medical literature, the focus has been on making these treatments affordable in low-resource countries (Ombelet and Campo, 2007; Ombelet and Onofre, 2019; Chiware et al, 2021).

In contrast, fertility professionals in Western countries have been reluctant to rethink the current standard approach (Cooke, 2016). For instance, professionals perceive the use of less expensive minimal ovarian stimulation protocols as risky because they result in fewer eggs per cycle and, therefore, are expected to have lower efficacy on a per cycle basis (Heng, 2007). Unsurprisingly, there is no conclusive evidence regarding the effectiveness of different stimulation protocols and several systematic reviews and meta-analyses have reached conflicting conclusions (Fan et al, 2017; Youssef et al, 2018; Datta et al, 2020). These reviews highlight that, due to the array of protocol options, there is a lack of consensus even on the definition of low or mild stimulation.

Promoters of these protocols emphasize that, in addition to lowering the costs for the patients and society, they reduce patient discomfort and risk of ovarian hyperstimulation syndrome (OHSS), therefore reducing dropout rates. As a result, some studies indicate that, when considering the cumulative live birth rate per woman, these 'lite' treatments can be successful in more than two-thirds of cases (Ferraretti et al, 2015; Gianaroli et al, 2022).

My intention here is not to endorse low or mild stimulation protocols, which can become a form of add-on themselves when misleadingly offered as 'natural' cycles to attract patients who resist the medicalization of reproduction.[15] Rather, my argument is that research into reducing the risk of stimulation protocols, and more broadly the cost of fertility care, has gained less attention from regulators, professionals, and patients because it is inconsistent with the dominant biomedical innovation model discussed in this book. Lower-dose stimulation protocols that promise equivalent or potentially lower chances of success, but with reduced costs and associated risks, are less attractive in a hope–driven market than expensive treatments that show *potential*. In fact, affordable solutions are explored and used mostly in countries where the high cost of fertility treatments renders them otherwise inaccessible; thus, they actually widen the fertility market. Additionally, these solutions are explored in places with legal or religious restrictions on creating more eggs and embryos. For example, the studies on 'lite' treatment mentioned earlier were conducted in Italy, where these protocols were developed during a period in which the national law prohibited the creation of more than three embryos per cycle (see Perrotta, 2013).

Conducting randomized controlled trials on the safety and efficacy of affordable fertility care should be a priority, as this would allow access to a large number of patients currently denied access worldwide and reduce dropout for financial reasons. As Marcia Inhorn (2015) has underlined, the high cost of fertility treatment has implied not only that these are not accessible in so-called developing countries and for many individuals in developed ones, but also that only few governments have been able and willing to fund these treatments worldwide. However, this path is unlikely to be taken in the hope market, unless new forms of regulatory interventions that aim at fostering responsible biomedical innovation are introduced. I will discuss some of these potential interventions in the Conclusion.

Conclusion: Fostering Responsible Innovation in Fertility Care

In this book, I argue that controversial fertility treatment add-ons should be framed as examples of a stabilized neoliberal model of biomedical innovation within a hope market. Throughout the book, I have developed the argument that this model succeeds by promoting and selling the *potential* of biomedical innovations, capitalizing on the uncertainties of infertility and medical knowledge. I use the term *hope market* to emphasize that not only do patients become customers responsible for purchasing their own treatments, but this market is also deeply shaped by hope logics, rather than purely economic ones.

Within a field characterized by high levels of uncertainty and widespread gaps in evidence, the allure of this potential is strong for patients, who often end up opting for and purchasing not just unproven but risky interventions. The numerous instances of past interventions, now with new evidence suggesting they might have actually hindered patients' chances of conceiving, shed light on the dark side of optimistic narratives surrounding potentiality.

Examining fertility care through the lens of a hope market prompts some concluding reflections on both the commercialization of health services and its regulation, within and beyond fertility care. The intricate landscape described in this book, where hope, medicine, and markets are interwoven, calls for a careful re-evaluation of practices and regulations to ensure that patients are not only well-informed but also protected from market dynamics that aim at maximizing profit. In the ensuing discussion, I will specifically address the risks associated with the uncritical promotion of informed choice within a hope market and contemplate the absence of discourse regarding corporate responsibilities in promoting unproven biomedical innovations. The strategies proposed, from altering reporting practices to holding private organizations accountable, seek to strike a balance between enabling patient autonomy and fostering responsible biomedical innovation. While this book does not delve into the specifics of regulatory interventions, I contend that such strategies should be integrated into the ongoing discourse surrounding regulatory changes in fertility care and add-ons, seeking to engage all pertinent stakeholders.

Unpacking informed choice

Patient choice is central in the experience of infertility. Extensive literature highlights the multitude of difficult decisions faced by individuals who are involuntarily childless. These choices encompass a range of options, including the socially challenging decision to refrain from taking any action at all (Strathern, 1990). The options presented to infertile individuals include, for instance, whether to start in vitro fertilization (IVF) treatment (Leyser-Whalen et al, 2018; Mounce et al, 2022) or consider less invasive options like intra-uterine insemination (Bahadur et al, 2020, 2021; Homburg, 2022). Patients also grapple with choosing a clinic (Marcus et al, 2005), determining whether to discontinue treatment (Peddie et al, 2005; Carson et al, 2021) or look for other alternatives to become parents, such as adoption and fostering (Daniluk and Hurtig-Mitchell, 2003; Peddie et al, 2004). As I have explored in this book and elsewhere (Perrotta and Hamper, 2021, 2023), with the proliferation of optional treatments, fertility patients have been confronted with additional choices pertaining to the components of their fertility treatment.

Several initiatives, such as the Human Fertilisation and Embryology Authority (HFEA) treatment guide for patients and the 'key questions to ask your clinic about treatment add-ons',[1] aim to facilitate informed choice by equipping patients with questions to ask clinics about various aspects of fertility treatment. These include add-ons, costs, and efficacy of interventions in the context of individual patients. Similar initiatives have been widely promoted within health care to assist patients in identifying tests, treatments, and procedures that may have uncertain or limited value. An example of this is the Choosing Wisely® campaign. Originally initiated by the American Board of Internal Medicine and subsequently adopted in the UK by the Academy of Medical Royal Colleges, this campaign serves as a means to aid patients in making well-informed decisions about their medical care by advocating for informed choices.

While these methods of empowering patients are designed to mitigate potential negative outcomes associated with excessive choice, particularly in terms of overuse and unnecessary care, they also come with certain limitations. Firstly, the dearth of evidence complicates the task of evaluating the responses to the questions patients are encouraged to raise. As I have thoroughly examined, most often professionals would not be capable of addressing questions regarding efficacy based on high-quality evidence, as this evidence is absent. Secondly, the questions posed can cast the interaction between patients and professionals in an atmosphere of distrust, positioning patients in a defensive stance against unwarranted offers. For instance, one of the recommended questions reads, 'Will you, or anyone employed by this clinic, benefit financially from me having this add-on?' An approach that relies on patients to assess whether clinics have vested financial interests may not only sour their experience but also breed distrust towards medical professionals and the healthcare system as a whole.

In broader terms, treating patients as consumers who bear responsibility for their choices, these approaches contribute to patient responsibilization, and subsequently, subjecting them to blame. As highlighted by some critics, the focus on the notion of choice within healthcare provision aligns with a broader shift of responsibility from governments to private individuals (Moss, 1998; Harvey, 2005). This trend is prominent in the current neoliberal political landscape, marked by the reduction of public provisions for health, education, and welfare services. In the healthcare sector, the emphasis on choice has promoted what has been defined 'health consumerism' (Greener, 2009), which blurs the lines between patients and consumers (Gabe et al, 2015). While neoliberal promoters have frequently advocated for choice-based policies to empower patients, apprehensions have been raised regarding the consequent commodification of healthcare implied by these policies (Fotaki, 2010, 2013).

In the highly commercialized field of fertility care, situated within a neoliberal framework characterized by a culture emphasizing self-actualization-through-choice (Salecl, 2011), it is not surprising that both patients and professionals interviewed in this research advocate for patients' ability to exercise informed choice about their treatment. In this context, informed choice is often conceptualized as a near-right, especially for patients who privately fund their treatment, highlighting the conflation of *patient* and *consumer* choice. As promoters of a free market of innovation in this field have underlined, patients/customers are not forced to purchase these treatments and the choice to invest in them should not be denied to those who can afford them, provided they are adequately informed. However, as I explored in Chapter Three, perspectives on the extent to which treatment options should be provided, and the potential unintended repercussions for patients, vary considerably.

As others have noted (Faircloth and Gürtin, 2018), while intended to further support those patients who want to direct their treatment, this superabundant level of choice generates new burdens and responsibilities that contribute to patients feeling overwhelmed by the experience. As I have documented elsewhere (Hamper and Perrotta, 2023), foreseeing the need to cover expensive fertility treatment greatly influences (prospective) patients' life decisions, including avoiding substantial expenditures, such as purchasing a new car or relocating to a larger, family-oriented residence, in order to save funds for IVF. Others must explore various funding options, such as relying on family financial support or even remortgaging their home (Fertility Network UK, 2023). Furthermore, the proliferation of treatment options has exacerbated an already unequal access to fertility care across the UK and worldwide. Yet, there remains limited understanding of the added burden on patients without the financial means to afford these treatments, intensifying feelings of responsibility for both their inability to conceive naturally and their inability to finance treatment.

Finally, the choice of optional interventions that *might help* to achieve the priceless promise of a baby is particularly problematic due to the dearth of conclusive evidence. As I extensively discussed in this book, widespread uncertainty and lack of knowledge mean patients cannot actually be *fully informed* about the benefits and risks of the interventions they are expected to choose. Indeed, in this context, being 'informed' often translates into understanding that the benefits and risks of these interventions are unknown. In addition to the imperative for greater transparency and access to comprehensive information, this underscores the necessity for redefining what constitutes informed choice within fertility care and it prompts a vital discussion on how to advocate for shared responsibility in the choices presented to patients.

Regulatory challenges

Regulators, professionals, and patients have raised several concerns regarding the rightness of selling unproven treatments and several strategies have been implemented or suggested to foster responsible innovation in this field.

A first strategy, encapsulated in the consensus statements on add-ons discussed earlier, envisions the use of these interventions in research only until conclusive evidence of their effectiveness is established through randomized controlled trails (RCTs). This strategy, which was supported by some professionals interviewed in this research, has been endorsed by a growing medical literature openly criticizing the hasty introduction of these interventions in clinical practice (Datta et al, 2015; Harper et al, 2017; Wilkinson, Malpas et al, 2019). For instance, the argument behind this position is convincingly explained by Sebastiaan Mastenbroek and colleagues (Mastenbroek et al, 2021) in a recent article aptly titled 'The imperative of responsible innovation in reproductive medicine'. Using the case of preimplantation genetic screening/pre-implantation genetic testing for aneuploidy (PGS/PGT-A) and its historical development, the authors illustrate how until novel interventions are not appropriately tested, it is not possible to determine not only whether they are effective, but also whether they could have negative effects. The case of PGS/PGT-A serves as exemplar, as numerous embryos previously classified as abnormal (mosaic) following genetic testing have been discarded over the years. However, as discussed in Chapter Two, current knowledge indicates that these embryos can actually result in successful pregnancies. As a result, it has become evident that many patients who invested in earlier versions of PGT-A may have unknowingly paid to decrease their chances of having a baby.

While there is consensus on the need to establish an evidence base for novel treatments, the regulation of offering these treatments before this is established remains contentious. Supporters of the consensus statement call for a ban on the market of unproven treatments due to the potential risk of actually reducing patients' chances (Harper et al, 2012, 2017; Repping, 2019; Wilkinson, Bhattacharya et al,

2019; Wilkinson, Malpas et al, 2019). Promotors of novel interventions believe that a tighter regulation would hinder innovation and would deny potentially beneficial treatment to generations of patients who have to wait for an evidence-base to be established (Cohen and Alikani, 2013; Murdoch, 2017; Macklon et al, 2019). Considering the challenges of evidence production in this field, as detailed in Chapter Two, it is not surprising that the implementation of a ban on unproven treatment has encountered resistance from various stakeholders.

As a countermeasure, regulatory efforts have focused on enhancing the provision and increasing the quality of information regarding these treatments. While these attempts are undeniably essential and laudable, strategies that focus on this approach present limitations. Firstly, these strategies do not adequately assist either patients who wish to place their trust in their healthcare providers or those who might succumb to the allure of potential benefits. More concerningly, as discussed in the previous section, they delegate the full responsibility of treatment decision-making to patients, leaving them even more accountable for making challenging decisions under the umbrella of informed choice promotion (Wilkinson, Malpas et al, 2019). The limited scope of what it means to be informed in this field, specifically grasping the absence of medical certainty regarding treatment effectiveness or safety, coupled with the prevalent direct-to-consumer marketing strategies promoting an optimistic portrayal of potential, undermines the effectiveness of these strategies in facilitating patients' decision-making.

To complement and enrich these strategies, I argue that questions regarding whether clinics have financial interests to sell specific interventions and whether they use them as income generators should be embedded in the regulatory framework. While implementing these strategies would not be possible under the current Human Fertilisation and Embryology Act 1990/2008, as the financial probity of the fertility sector is not currently under its remit, these should be considered as part of the current review of the law. For instance, incorporating the use of certain interventions along with their associated financial information in national reports would require limited effort, yet offer invaluable data to assess the existing scenario. As discussed earlier, despite the controversy and the regulatory interventions, very little is known regarding the value of this market. Providing financial information about the expenses associated with treatment and different interventions would serve as a valuable tool for patients' decision-making, as this entails both medical and financial aspects. Moreover, these data could also be employed to enhance the HFEA guidelines for selecting a clinic.[2]

In an open economy such as Britain, this type of reporting would align with broader principles of corporate transparency and previous parliamentary discussions on a bill on fertility treatment transparency,[3] which has a specific focus on add-on treatments. However, this type of approach should be considered on a global scale as a more favourable option than contemplating a ban on unproven innovations in a field where it remains difficult to determine whether most treatments are supported by high-quality evidence. In addition, such a shift in reporting practices would complement recent appeals to modify the reporting of success rates, as the existing emphasis on live birth rate per embryo transfer is deemed to encourage the premature adoption of these interventions (Fauser, 2019).

Moreover, regulatory support could facilitate the implementation of a series of smaller-scale initiatives, including establishing confirmed average price lists to enable patients' comparisons, offer incentives or penalties for clinics based on information accuracy, and oversight the compliance of clinic websites to information standards.

Standardizing treatment prices in the fertility market seems unfeasible due to significant differences in expenses incurred by clinics, such as rent for local premises or consultant fees varying by seniority (Perrotta and Smietana, 2024a). However, the lack of standardization of fertility treatment costs across the sector hinders patients' ability to compare. Some recent initiatives, such as the Fertility Mapper,[4] provide a space for patients to report the cost of clinics. However, these are based on self-reporting and are heterogeneous. Regulatory support to produce a trusted and reliable list of average national prices per treatment would enable patient comparison.

As part of their call to modernize fertility law, the HFEA is advocating for expanded powers to regulate the commercial aspects of the field, including the authority to fine clinics for mis-selling add-ons. However, due to the uncertainty generated by the lack of evidence, it is challenging to envision how such a system would operate in practice. Nevertheless, implementing measures to assess the accuracy of information based on current Competition and Market Authority (CMA) guidelines – such as verifying clinics' compliance with the expectation to provide evidence for benefit statements, report risks, and link to the HFEA rating system – and associating them with incentives or penalties could be a potentially useful measure to implement.

Despite existing guidelines and assessment procedures for clinic websites, the provision of clear, transparent, and up-to-date information remains unguarded (Perrotta and Smietana, 2024c). Ensuring clinics comply with information standards is essential to enhance the overall accuracy and currency of information provided. For example, UK fertility clinics are legally mandated to disclose their success rates and costs on their websites, with obligations under consumer law. Despite oversight from both the Care Quality Commission (CQC) and HFEA inspections, as discussed earlier, the information on clinic websites appears inconsistent. While HFEA inspections have broader aims, oversighting the compliance of clinic websites to information standards could be implemented with the support of engaged stakeholders, such as professional bodies and patient organizations.

While each of these small-scale measures presents its own set of challenges and considerations, collectively they signify potential steps toward enhancing transparency, accountability, and patient empowerment within the fertility market.

Corporate responsibilities within biomedical innovation

The ongoing debate surrounding add-ons, discussions about potential regulatory shifts, and the diverse array of literature on fertility care have surprisingly neglected to explore the types of corporate governance prevalent within this sector. As some of the professionals interviewed in this research have emphasized, fertility clinics which are private businesses need to produce profits to remain in the market. However, the array of commercial practices scrutinized in earlier sections prompts a discussion on the standards that should be expected from private companies operating in the healthcare sector.

In the UK, while paying regard to other stakeholders rather than adopting a narrow financial perspective, company directors have a legal duty to promote the success of the company for the benefit of its members (shareholders) as a whole. In the current corporatization of the fertility sector, clinics owned by private equity firms are managed by non-medical professional managers (Patrizio et al, 2022), fertility professionals are called to develop expertise in business and management (Masler and Strickland, 2013; McLaughlin et al, 2019) and even develop financial fluency to maintain successful practices (Christianson et al, 2021). However, the literature has neglected any discussion of potential voluntary practices that might have a positive impact on a wider range of stakeholders, including (prospective) patients and the wider society, or their regulatory enforcement.

This complete lack of discussion is paradoxical in a sector that provides healthcare and a vital service such as fertility care. My contention is that these forms of regulation have not been contemplated because the attention has been placed on specific add-ons rather than on the biomedical model that supports their proliferation. Here, I call for a complete reversal of viewpoint. Instead of investigating further approaches that place the responsibility on patients for purchasing unproven treatments, alternative strategies should be put into action to hold health organizations accountable for marketing and selling such interventions.

For instance, in other medical fields and healthcare services there is an ongoing discussion of voluntary forms of corporate social responsibility (CSR). CSR refers to practices that 'further some social good, beyond the interests of the firm and that which is required by law' (McWilliams and Siegel, 2001, p 117). Through CSR practices, health organizations are expected not only to abide by the law and follow general ethical principles but also contribute positively to society, minimize negative impacts, and promote sustainable practices. These practices should take into account the interests of all their stakeholders, including (prospective) patients and wider society (Brandão et al, 2013; Tehemar, 2015; Haddiya et al, 2020).

The interests of patients should be among the top priorities for any organization wishing to operate in the field of fertility care. Not only because this is ethically and morally just, but also because a problematic model of biomedical innovation such as the one I have discussed in this book entails a variety of reputational risks. These include potential legal costs arising from threats of litigation (Manning, 2018), the possibility of losing patients due to a tarnished reputation (Barnett, 2023; Chawla, 2023), and a wider erosion of trust in the private fertility sector.

Additionally, in recent years, several frameworks have emerged to address the tensions between companies' financial objectives and broader societal concerns. For instance, the concept of *purposeful business* emphasizes that a company's goal should be to address the problems of people and planet profitably, rather than profiting from causing them (British Academy, 2019). This approach recognizes that competition alone may not align business interests with those of society, necessitating a substantial regulatory overhaul that encourages companies engaged in significant social functions, like fertility care, to integrate public purposes into their corporate missions.

As I discussed elsewhere (Perrotta and Smietana, 2024b), a common notion in the sector is that without charging patients for innovative techniques, there would be insufficient resources to generate evidence over the long term. While this perspective holds merit, as I have shown throughout the book, only a dearth of overall evidence has been produced. Rather, new iterations of existing techniques or treatments are introduced, accompanied by claims of newfound potential. Similarly, older tools are rebranded as cutting-edge technology, as exemplified by the current trend of offering time-lapse imaging (TLI) under the guise of artificial intelligence (AI).

I argue that the focus in fertility care should not be on banning innovation but rather on ethically regulating its market. As part of their corporate social responsibility, clinics should adopt the ethical imperative of prioritizing patient well-being over financial gains and refrain from profiting from unproven interventions. Instead, any profits generated should be allocated towards funding the establishment of an evidence base before offering certain treatments.

Beyond fertility care

In this book I illustrate the dynamics behind biomedical innovation in the field of fertility care. My core argument is that the lack of regulation of this hope market allows private organizations, very often private equity firms, to capitalize on the uncertainties of the experience of infertility and of medical knowledge, creating hope markets. In these markets patients, who are made to hold the responsibility of their treatment through a blurred notion of informed choice, are turned into customers who purchase the potential of unproven and often risky biomedical innovations. Such a model exacerbates inequality among those who can afford these innovations and those who cannot, but also embitters the experience of many who either feel compelled to get into debt to pay for these treatments or feel further responsible for not being able to pay to achieve their desired child. While this model of biomedical innovation works well in terms of generating funding for companies who manufacture or practise these interventions, this does not serve either patients or the wider community which fund innovation either by purchasing them directly or through public national health systems. In fact, some recent studies suggest that over the last decades, and in parallel with the establishment of this model of biomedical innovation in this field (Patrizio et al, 2022), fertility treatment success rates have actually decreased (Gleicher et al, 2019).

While the case I have explored is based in the field of fertility care and located in a specific country, the UK, I argue that this model is not specific to either the medical sub-field or the nation. While I am not aware of other studies looking at biomedical innovation as I have done in this book, similar concerns as those addressed here are increasingly emerging from different strands of literature. Concerns regarding the effect of privatization have been raised in various branches of medicine. These concerns are prevalent in those sub-fields, such as dermatology, cosmetic treatments, and dental care, that do not prevent death but promote health, and are, therefore, considered unnecessary and poorly publicly funded. Moreover, these concerns are widespread in all areas of medicine in those countries where medical practice is completely privatized, as in the USA (Borsa et al, 2023). The combination of a demand that is determined by factors that are not only or primarily related to the cost of these products and aggressive commercial logics that aim to expand these markets and generate new needs fosters further medicalization of life stages. Primary examples of these are the proliferation of products for the management of menopause or aging (Cardona, 2008; MacGregor et al, 2018; Geddes, 2022; Larocca, 2022).

As I have detailed in this book through the case of fertility care, the inability of evidence-based medicine (EBM) to fill its promise to resolve medical uncertainty has a central role in the spread of such a model of biomedical innovation. Producing certainty through EBM requires not just substantial funding to perform large RCTs, but also the skills and expertise to produce and interpret evidence. Paradoxically, very often, those who are able and willing to fund these RCTs are also those who have a vested interest in selling the products that these RCTs are testing (Bartels, 2009). The recent wave of concerning studies (Every-Palmer and Howick, 2014; Ioannidis, 2016; Epstein and ProPublica, 2017) which denounce the use of manipulated or entirely fabricated evidence used as a marketing tool to sell biomedical interventions shows how EBM alone cannot protect patients/customers in hope markets.

The model of biomedical innovation described in this book is optimal only for the few actors benefitting from the revenues that it produces, while it is highly detrimental for everyone else (Trayer et al, 2022). These include (prospective) patients who are pushed to pay for the potential of medical innovation and, de facto, fund it, while being blamed for purchasing unproven and potentially hazardous treatment. However, it also includes medical professionals, who are often deemed responsible for over-treating and over-prescribing in a context that equally pushes them into these practices, and medicine as an institution, as trust in medicine is being eroded by the consequences of this model (Jureidini and McHenry, 2020). Finally, it includes wider social communities worldwide. On the one side, because the examples of biomedical innovations (as discussed in Chapter Two) that are widely used although they do not help or are not needed extend across all areas of medicine. And on the other hand, because the cost of this biomedical model of innovation in terms of public health is disproportionate, as US-based studies have already shown (Patashnik et al, 2020).

The issues raised in this book are not specific to the fertility industry and further research looking at the dynamics described here is urgently needed. However, as I have articulated in this section, there are clear indications that this model is spreading worldwide and is already fully established in some countries and some areas of medicine. My final contention is that this is due to a combination of the lack of public funds for medical health and the unregulated privatization of healthcare witnessed in many Western countries.

As Debora Spar (2010) highlights, markets in health services do not work as well as expected. Matching supply and demand through price fluctuation mechanisms should increase supply over time and reduce price, providing greater access to any commercially available product. However, as I have shown in this book, the hope market does not follow these rules, as social norms and future life expectations make these products invaluable and are inherently exploitative. While neoliberalism supporters claim that regulating private markets would hinder innovation, the case of fertility care shows that fostering innovation that does not work does not benefit anyone.

I argue that reconsidering the regulatory framework to redistribute responsibility towards health organizations, by regulating the organizational forms through which medical and health services are provided, would yield better outcomes for all stakeholders.

Appendix: Methodology Notes

This book is mostly based on data collected through a Wellcome Trust-funded project titled 'Remaking the Human Body: Biomedical Imaging Technologies, Professional and Lay Visions' (grant no. 108577/Z/15/Z), which I led between 2016 and 2023. As the title suggests, when in 2015 I applied for funding for this project to explore the introduction of time-lapse incubators in fertility care, I was expecting to investigate mainly the relations among new visual technologies and professional and lay visions. The abrupt advent of add-ons and the subsequent change in the public and medical discourse on time-lapse as one of them required some amendments to the original methodological design of the project, which included ethnographic observations in fertility clinics and dedicated events, as well as individual and group interviews with fertility professionals, patients and other relevant stakeholders, as detailed in Table A.1.

While the original design has withstood the add-ons storm, both the interview guides and the content of the interactions with various actors throughout the project have been significantly shaped by the abrupt change of narrative. In just a few months, from revolutionary technologies with the potential to change embryology and fertility care, time-lapse incubators became an unproven technology used to exploit unaware patients.

Despite seeking to conduct observations in both private and National Health Service (NHS) clinics, I have not succeeded in obtaining access to private clinics, excluding a day visit to one clinic and tour of its laboratories conducted by Dr Anne-Sophie Giraud after an interview. Instead, many of the NHS clinics where observations were conducted approached me to

Table A.1: Data collected through the Remaking the Human Body project

	NHS clinics	Private clinics	Other stakeholders
Ethnographic observations	5 clinics (230 hours of observation)	1 clinic (6 hours of observation) 6 open evenings (observation) 2 informal discussions	Fertility events The Fertility and ESHRE conferences, Fertility Fest, Fertility Show, PET events
Interviews with professionals	8 semi-structured interviews	25 semi-structured interviews	10 semi-structured interviews
	4 heads of embryology	2 heads of embryology	4 company representatives
	14 embryologists	2 embryologists	2 professional body representatives
	2 research nurses	2 clinical directors	2 evidence/trials
	2 nurses	1 doctor	1 bioethicist
	1 clinical director	1 andrologist	1 patient advocate
	2 doctors		
Interviews with patients	50 semi-structured interviews (51 participants: 41 female patients, 9 male partners and 1 female partner) Aged between 29 and 47 (mean age of 36) Mix of NHS- and privately-funded treatment Recruited via the collaborating NHS clinics and an online survey		
Focus group interviews	6 focus group interviews (35 participants) 3 with professionals and 3 with patients (5–7 participants each)		
Online survey	314 patients		

participate in the research through the National Institute of Health Research Clinical Research Network. The eligibility criteria for clinics to join the project was to use a time-lapse incubator (of any brand or model), but, as I discuss throughout the book, several among them offered other interventions labelled as add-ons to some patients.

While Table A.1 presents data according to the type of organizations where professionals were working at the moment of interviews, the distinction between NHS and private care is ambiguous for both professionals and patients. Excluding a few cases in which senior professionals had built their careers (and identity) in the public or private sectors, most of the professionals we interviewed had had experience in both: some because they had worked in different clinics during their career, others because the clinic where they were working had gone through changes in the management and organizational structure. Similarly, as I have examined elsewhere (Hamper and Perrotta, 2023), patients' experiences are hybrid. Many had received both NHS- and privately-funded treatment, because they had not been successful during the NHS-funded treatment(s) they were entitled to and/or they were trying for a second child. In addition, expectations regarding the need to consider funding their own treatment, often even before starting it, shaped patients' views and experiences (Hamper and Perrotta, 2023).

The research was conducted in different stages, and for each I secured ethics approval through the appropriate process. Observations and interviews with professionals and other stakeholders working in the private sectors were approved by the Queen Mary University of London Research Ethics Committee (QMERC 2015.80a.30.11.16 and 2015.80b.30.03.17). Observations conducted in NHS clinics in England, interviews with professionals working in the NHS, and interviews with patients were approved by the Health Research Authority (REC Reference 17/EM/0218) and each clinic site separately. When clinics agreed to participate in the research, all staff were informed ahead of time about study procedures.

Ethnographic observations in clinics, carried out by Dr Alina Geampana and myself, involved observing laboratory and clinical practices (including gamete and embryo manipulation, egg collection and embryo transfer), as well as staff interactions in relation to the conduct of clinical trials. In addition, we shadowed embryologists during their daily laboratory routines, use of time-lapse tools, and interactions with other clinic staff. Together with members of the research team, I also conducted observations in a variety of fertility events, including regular participation in the Fertility Show, attending professional conferences, clinic open evenings and events organized by the Progress Educational Trust.

During observations in clinics, Dr Geampana and I directly contacted professionals at the observation sites during the observations to invite them for interviews. Participation was voluntary, and interviewees provided additional consent by signing a form. In total, 25 interviews were conducted with NHS staff, comprising embryologists, consultants, nurses, and clinic directors. We also recruited an additional 18 key stakeholders via email based on their expertise, including clinicians, researchers, and representatives of professional bodies. All interviews with professionals were conducted between June 2017 and July 2019, either face-to-face (35 interviews) or over Skype or phone (eight interviews) where meeting in person was not feasible. These lasted between 35 and 100 minutes and were recorded and professionally transcribed verbatim.[1] Our interview guide focused on obtaining participants' experiences with using time-lapse and add-ons, conducting clinical trials, and opinions on new and controversial add-ons as well as issues with producing reliable evidence in the fertility care sector. We aimed to collect responses without expressing any personal views on evidence and evidence-based medicine (EBM) practice in fertility care.

Data collection also included interviews with 51 fertility patients: 40 women who were going through treatment, 10 male partners and one female partner. Most of these interviews were conducted by Dr Josie Hamper, while a group of patients was interviewed by Dr Geampana. Patients and their partners were recruited using two routes to ensure a diverse representation of those undergoing NHS-funded, privately-funded, or combined treatments. Initially, 31 participants were enlisted by research nurses from three of NHS fertility clinics participating in the research, one based in London and the other two across England. Subsequently, 20 participants were engaged through an online survey (N = 314) distributed on relevant social media platforms. As the survey was created as a screening tool to recruit patients that had experienced time-lapse and other add-ons in the private sector, this collected further information about the variety of treatments used by patients. The survey allowed participants to provide their contact information for potential further research involvement. Participants recruited through NHS clinics were interviewed on-site at these clinics, while the other interviews were conducted via phone, at participants' homes, or in café settings, based on individual preferences.

Every interviewed patient and partner received written details about the study and had the chance to seek clarifications before granting written consent to take part in the research. Each interviewee was offered a £20 voucher for their participation. The interviews ranged from 30 to 80 minutes in duration, accumulating approximately 40 hours of recorded interviews, which were then professionally transcribed verbatim. At the time of the interviews, the age range of these participants was 29 to 47, with an average age of 36. Patients and partners were at different stages of their fertility journey: 26 of the participants were either undergoing IVF treatment or preparing to start it; 15 were either pregnant themselves or had partners who were pregnant; and 10 had already experienced successful IVF cycles resulting in one or more children. While both recruitment routes aimed to include both men and women, recruiting male participants among this group presented challenges, reflecting a well-documented difficulty in social scientific research on (in) fertility (Culley et al, 2013). Generally, these participants came from socially or economically privileged backgrounds, which likely facilitated their access to fertility treatment. However, our participant group was diverse in terms of their ability to afford private IVF treatment.

In addition to individual interviews, we conducted six focus group interviews, three with patients and three with professionals. Each focus group had between five and seven participants, with a total of 35 participants. The three focus groups with professionals and one of those with patients were recruited via the three of the NHS clinics participating in the research and conducted on clinics' premises between 2019 and 2020. Due to the restrictions imposed by the pandemic, the other two focus groups with patients were conducted online in 2021. Participants in these last focus groups were recruited via the screening survey. As for the individual interviews, most participants in the patients focus groups were female. Three men participated in these focus group together with their female partners. As in the case of individual interviews, participants were at different stages of their fertility journey.

While some of the professionals who participated in the focus group had already been interviewed individually, only one of the female patients participated in both. Written consent was obtained following the same procedure of the individual interviews. Patients and partners were offered a £20 voucher for their participation, while the participation of professionals was on a voluntary basis. We developed a novel methodology for the focus groups. We used two short videos (a clip from a BBC documentary on infertility and a clinic advertisement of time-lapse), and six excerpts from interviews (three from professionals and three from patients) re-enacted by actors to elicit group discussion (see Geampana and Perrotta, 2024). We used the same interview guide for the focus groups with professionals and patients to investigate reciprocal understanding of the other group's logics and rationales. All six focus group interviews lasted approximately 90 minutes and were then professionally transcribed verbatim.

During the project, members of the research team and I performed a variety of data analyses focusing on specific aspects. The results of these analyses have been included in several articles published in social scientific journals, which I refer to throughout the book when appropriate. The reflections presented in this book have been further developed through a round of situational analysis (Clarke, 2003) which I performed myself in 2023. This further analysis was supported by the British Academy through an Innovation Fellowship (grant no. IF2223\230087), and included a variety of additional documentary materials. Following the tenets of situational analysis, I created a situational map that sought to include the full range of positions on add-ons to situate my interpretation and elaborate an ecological understanding of the complexity of the add-ons case. To do so, I had to integrate the data previously coded with additional materials able to represent the missing viewpoints. These materials included a wide variety of public documents, such as documents and minutes of the Human Fertilisation and Embryology Authority (HFEA) meetings published on their website, quantitative data on the use of add-ons from various sources, and a vast number of medical guidelines, documents, and reports.

This analytical approach allowed me to integrate this variety of viewpoints and generate the sensitizing concept of *hope market*, and articulate what was missing from the discussion on add-ons: a discussion on the model of biomedical innovation that these optional interventions embed. Most of the discussions, both in the public debate and in the data collected, framed this case as a controversy between those supporting the risk of innovation and those privileging the need to establish evidence first. However, such a dichotomy is not generative, as it reinforces the standpoints of different groups and the values that these views encapsulate (for a discussion, see Geampana and Perrotta, 2022). The neoliberal model underpinning these opposing positions was never acknowledged, and therefore not even challenged. While the novel interpretations discussed in this book are an empirically grounded theoretical integration, I hope that the policy suggestions that they originate might improve the experience of all the actors involved.

Notes

Introduction: Biomedical Innovation in Fertility Care

[1] See: 'Couples exploited by fertility clinics offering "add-ons"', *The Times*, 2016, available at https://www.thetimes.co.uk/article/couples-exploited-by-fertility-clinics-offering-add-ons-wq8qzwdph (last accessed 20 April 2024); 'How IVF became a licence to print money', *The Guardian,* 2018, available at https://www.theguardian.com/lifeandstyle/2018/jun/18/how-ivf-became-a-licence-to-print-money (last accessed 20 April 2024); 'Scandal of the risky rip-off IVF "add-ons"', *Daily Mail,* 2020, available at https://www.dailymail.co.uk/news/article-9091465/Scandal-risky-rip-IVF-add-ons.html (last accessed 20 April 2024).

[2] Some of the ideas presented in this book initially took shape during my previous work leading the Wellcome Trust-funded project 'Remaking the Human Body: Biomedical Imaging Technologies, Professional, and Lay Visions' (2016–23). The outcomes of the project have been published in several articles co-authored with collaborators who worked on the project. I am especially grateful to two former postdoctoral researchers, Dr Alina Geampana and Dr Josie Hamper, for their invaluable contributions to the project. Throughout the book, I refer to these articles for a detailed discussion of specific issues. However, in this work, I have synthesized and integrated these ideas to both theorize and document the challenges of the biomedical innovation model in fertility care, and I take full responsibility for the resulting content.

[3] As part of the growing field of ignorance studies (Gross and McGoey, 2015; McGoey, 2016), recent social science literature in the broader field of reproductive medicine has examined ignorance production as a lens to understand risk and innovation (Bell, 2022; Topçu and Maffi, 2022). While scholars in this field highlight the regularity and even orderliness of using knowledge gaps in decision-making processes (McGoey, 2014; Knudsen and Kishik, 2022), I prefer to avoid the negative connotation of the term ignorance and instead solely use the terms uncertainty and 'nonknowledge'.

[4] Other pregnancies had been previously reported by the team and other teams worldwide, but either they did not continue, or they were ectopic and had to be terminated to safeguard prospective mothers (Cohen, 2012).

[5] While I recognize the importance of addressing forms of involuntary childlessness experienced by single individuals and same-sex couples,

this book's focus is primarily on the broader trajectory of biomedical innovation in fertility care. For an in-depth discussion of the consequences of certain definitions of infertility for specific groups of people, I refer readers to Cavaliere (2023).

[6] The data from HFEA (2023) for the year 2021, indicating an average embryo transfer success rate of 22 per cent, is preliminary at the moment of writing. However, this figure is anticipated to rise after data validation, considering that birth rates generally lag three percentage points behind pregnancy rates, which currently stand at 29 per cent.

[7] Data for Table 0.1 and 0.2 is from the HFEA Fertility Trend Report 2021, which is available at https://www.hfea.gov.uk/about-us/publications/research-and-data/fertility-treatment-2021-preliminary-trends-and-figures/ (last accessed 20 April 2024). Specifically, data reported in Table 0.1 comes from section 1.

[8] Information regarding the funding system in the UK is available on the HFEA website at https://www.hfea.gov.uk/treatments/explore-all-treatments/costs-and-funding/ (last accessed 20 April 2024). Details on the criteria for funding in each region of the UK are available on the Fertility Network UK webpage at https://fertilitynetworkuk.org/access-support/nhs-funding/ (last accessed 31 July 2023).

[9] Adopting the criteria of identifying NHS clinics by their display of the NHS logo or hosting their website on an NHS URL led to 12 incorrect identifications.

[10] Though the quality and accuracy of these estimates and predictions may be subject to debate, it is noteworthy that several major market research organizations, such as Insight10, Market Research, IBISWorld, and Allied Market Research, provide market research reports for the fertility care market in the UK.

[11] In the last two decades, several social scientists have explored various aspects of the global reproductive market, including gamete commodification (Almeling, 2007, 2011; Pfeffer, 2011; Krawiec, 2023), global commercial surrogacy (Pande, 2010, 2014; Hochschild, 2011; Markens, 2012) and, more recently, elective egg freezing (van de Wiel, 2020a; Gürtin and Tiemann, 2021; Takhar, 2023). Interestingly, while these practices have garnered significant attention in the literature, they constitute only a small fraction of fertility care. For instance, in the UK egg freezing and cycles using donor gametes (including egg, sperm or both) represent respectively less than 5 per cent and 12 per cent of the total number of cycles (HFEA, 2023).

[12] Marketing research has largely overlooked how fertility treatments are offered, with only a few notable exceptions (Houston, 2004; Takhar and Pemberton, 2020; Takhar and Houston, 2021; Takhar, 2023). While my

analysis is rooted in medical sociology and STS perspectives, I align with these scholars in advocating for expanded research within this area.

one What Are Fertility Treatment Add-ons?

[1] Information on the add-on rating system can be found on the HFEA webpage titled 'Treatment add-ons with limited evidence', available at https://www.hfea.gov.uk/treatments/treatment-add-ons/ (last accessed on 31 January 2024).

[2] More information about the evidence used to inform this decision can be found at https://www.hfea.gov.uk/treatments/treatment-add-ons/pre-implantation-genetic-testing-for-aneuploidy-pgt-a/ (last accessed 10 July 2023).

[3] The colours are now also associated with symbols (exclamation mark for red, plus–minus sign for amber, plus for green, question mark for grey and the symbol Ø for black).

[4] This statement was previously included in the HFEA traffic light system webpage under the section 'What do the traffic light ratings mean?', available at https://www.hfea.gov.uk/treatments/treatment-add-ons/#add-ons (last accessed 10 July 2023).

[5] The SCAAC, which is responsible for reviewing the traffic light system, acknowledged the uncertainty that arose from the previous amber rating for add-ons. In response, the committee has introduced a new grey category, acknowledging that evidence for treatment add-ons often presents conflicting information. All the meeting minutes and papers from each SCAAC meeting are available at https://www.hfea.gov.uk/about-us/our-authority-committees-and-panels/scientific-and-clinical-advances-advisory-committee-scaac/ (last accessed 10 July 2023).

[6] In Chapter Four, I will delve into the concerns raised by scholars (Kamath et al, 2019) regarding the increasing number of add-ons offered to patients. However, it is important to note that for most professionals, many of these treatments are considered to be a standard part of fertility practice in specific circumstances.

two Evidence Challenges in Fertility Care

[1] While I am indebted to the extensive body of social science literature that challenges the epistemological foundations of EBM, discussing this aspect falls outside the scope of this book. For more comprehensive insights into how this literature has influenced my understanding, please refer to my previous work (Perrotta and Geampana, 2020, 2021; Perrotta and Hamper, 2023).

2   I have discussed additional concerns regarding EBM in both medical and social science contexts elsewhere (Perrotta and Geampana, 2020). An extended review of the social science literature on EBM can be found in Stevens (2018).

3   The authors included all the Cochrane reviews that contained the term 'assisted reproductive technology'.

4   This method is also used when a patient is at risk of developing ovarian hyperstimulation syndrome (OHSS), a potentially serious condition that may arise from controlled ovarian stimulation during fertility treatments. However, it is considered an add-on when offered with the purpose of increasing the chances of success.

5   While the details regarding different TLI tools and algorithms go beyond the scope of this book, I have extensively explored this in my previous work. For an in-depth discussion of different models and their implications, please see Perrotta (forthcoming).

6   I am refraining from including direct quotes from the participants concerning this criticism to ensure the anonymity of the fertility professionals under scrutiny, as the matter extends beyond individual cases.

7   The updated version of the consensus statement can be found on https:// www.hfea.gov.uk/media/kublgcp3/2023-10-19-treatment-add-ons-consensus-statement.pdf (last accessed 25 January 2024).

8   We identified four distinct approaches to evidence interpretation in IVF: (1) delegating evaluations of evidence to experts; (2) critically assessing available evidence; (3) acknowledging the process of making evidence; and (4) contextualizing evidence in patients' lived experience of infertility. I refer readers to this work for details and nuances on each of these approaches.

three The Fertility Market: Help, Hype, and Hope

1   The Fertility Show is a biannual UK exhibition that brings together clinics, diverse experts and individuals interested in learning about various aspects of fertility treatment, options, and related topics. The exhibition features seminars, workshops, informational sessions, and opportunities for attendees to connect with experts and gather information about fertility treatments, adoption, surrogacy, and other related subjects.

2   Mumsnet is a popular online platform and community in the UK that provides a space for (prospective) parents to connect, share advice, and discuss various topics related to parenting, family life, and other aspects of daily living. The website features discussion forums, articles, and resources on a wide range of subjects, including trying to conceive and

infertility. Mumsnet has gained prominence as a place for people to seek and offer support, exchange experiences, and engage in conversations about a variety of topics.

[3] This section of the website is available at https://www.vitrolife.com/ivf-journey/ (last accessed 17 August 2023).

[4] This section of the website is available at https://www.vitrolife.com/ivf-journey/evaluation/ (last accessed 17 August 2023).

[5] This section of the website is available at https://www.vitrolife.com/ivf-journey/evaluation/time-lapse (last accessed 17 August 2023).

[6] This section of the website is available at https://www.vitrolife.com/ivf-journey/embryo-transfer/ (last accessed 17 August 2023).

[7] I acknowledge the extensive feminist literature that has illuminated how the aspiration for a biologically related child is imbued with gendered and normative implications and how fertility treatment can serve as a means to validate one's state of childlessness (see Throsby, 2010; Morris, 2019; Carson et al, 2021). However, this discussion exceeds the scope of this book.

[8] I have extensively examined various approaches to evidence in my previous work (Perrotta and Hamper, 2023), and I encourage readers to refer to it for a more comprehensive analysis.

four Regulating the Hope Market

[1] While a comprehensive review of various regulatory systems is beyond the scope of this book, it is essential to highlight the efforts made to harmonize these systems. The World Health Organization (WHO, 2017) has introduced a Global Model Regulatory Framework for Medical Devices, encompassing in vitro diagnostic medical devices as well. This initiative aims to establish a unified approach to regulating medical devices internationally, promoting safety and efficacy standards across different regions.

[2] Not all treatments fall under the category of medicinal products according to the MHRA. The MHRA assesses each product on a case-by-case basis to determine its classification as a medicinal product or otherwise. For more detailed information, please refer to the MHRA Guidance Note 8, 'A guide to what is a medicinal product', available at https://assets.publishing.service.gov.uk/government/uploads/system/uploads/attachment_data/file/872742/GN8_FINAL_10_03_2020__combined_.pdf (last accessed on 20 July 2023).

[3] It is essential to note that, as reported on the MHRA website, 'A medicine that is granted a licence does not necessarily become immediately available to patients in the UK. The Department of Health and Social Care make

decisions on which medicines should be purchased for the UK, and the National Institute of Clinical Excellence decide whether or not medicines should be made available on the NHS. These organisations are able to consider wider factors in their decision-making, including the need for the medicine in the UK given the circumstances at that time.' Further information is available at https://www.gov.uk/government/publicati ons/more-information-about-the-mhra/more-information-about-the-mhra--2#medicines-and-vaccines (last accessed on 20 July 2023).

4   This does not apply to products registered as traditional herbal medicines or homoeopathic medicines, which are required to meet statutory standards of safety and quality only.

5   The latest guidance on the regulation of in vitro diagnostic medical devices in Great Britain, published by the MHRA in July 2023, is available at https://assets.publishing.service.gov.uk/government/uploads/system/ uploads/attachment_data/file/1173570/Guidance_on_the_regulation_ of_IVD_medical_devices_in_GB.pdf (last accessed on 20 July 2023).

6   After Brexit, there have been several changes to the essential requirements for placing a medical device on the Great Britain market (England, Wales, and Scotland – Northern Ireland follows EU regulations). Nevertheless, the current route to market in Great Britain and the UKCA marking requirements are still rooted in the regulations derived from the EU Medical Devices Regulation and EU In Vitro Diagnostic Medical Devices Regulation. These regulations can be accessed respectively at https:// eumdr.com/ and https://euivdr.com/ (last accessed on 20 July 2023).

7   For higher-risk devices (Class II, IIb, or III), the certificate must be issued by an independent 'UK approved body'. In contrast, lower-risk devices (Class I) do not require assessment by a UK approved body, and the manufacturer can self-certify the product before placing it on the market.

8   The SCAAC paper discussing the embryo culture media of February 2021 can be accessed at https://www.hfea.gov.uk/media/3317/scaac-embryo-culture-media-february-2021.pdf (last accessed on 20 July 2023).

9   Additional information on this proposal can be found on the HFEA's website at https://www.hfea.gov.uk/about-us/modernising-the-regulat ion-of-fertility-treatment-and-research-involving-human-embryos/ (last accessed on 20 August 2023).

10   The HFEA's press release regarding the consultation can be accessed at https://www.hfea.gov.uk/about-us/news-and-press-releases/2023-news-and-press-releases/fertility-law-needs-modernising-says-uk-regulator/ (last accessed on 20 August 2023).

11   The 9th edition of the Code of Practice (revised in October 2021) is available at https://portal.hfea.gov.uk/media/it1n3vpo/2022-07-01-code-of-practice-2021.pdf (last accessed on 20 July 2023).

12 Additional information can be found on the HFEA's website at https://www.hfea.gov.uk/about-us/how-we-regulate/ (last accessed on 20 July 2023).

13 This information is included in the HFEA clinic focus – July 2015, which can be accessed at https://ifqlive.blob.core.windows.net/umbraco-portal/1205/medicines-management-guide-for-clinics.pdf (last accessed on 20 July 2023).

14 In the past, embryos were typically transferred to the uterus on day three. Extended culture to day five is currently recommended when there are sufficient embryos on day three, as it allows for better selection of embryos for transfer.

15 The HFEA offers information about less medicated forms of treatment on their website, which is available at https://www.hfea.gov.uk/treatments/explore-all-treatments/ivf-options/ (last accessed on 20 July 2023).

Conclusion: Fostering Responsible Innovation in Fertility Care

1 The HFEA provides a list of 'key questions to ask your clinic about treatment add-ons' on their website. The complete list is available at https://www.hfea.gov.uk/choose-a-clinic/preparing-for-your-clinic-appointment/ (last accessed 20 August 2023).

2 The HFEA offers comprehensive information regarding the factors to take into account when selecting a clinic. This information can be accessed at https://www.hfea.gov.uk/choose-a-clinic/finding-the-best-fertility-clinic-for-you/ (last accessed 20 August 2023).

3 More information about the Fertility Treatment (Transparency) proposal (Volume 726: debated on Wednesday 18 January 2023) are available at https://hansard.parliament.uk/commons/2023-01-18/debates/BD41DF93-3E19-43D8-802D-CEFA50892778/FertilityTreatment(Transparency) (last accessed 20 August 2023).

4 Available at https://fertilitymapper.com/ (last accessed 20 January 2024).

Appendix: Methodology Notes

1 All the interviews were meticulously transcribed by an experienced professional using a 'denaturalized' transcription approach, employing the full verbatim style. This method entails retaining everything present in the spoken discourse, including utterances, repetitions, and all grammatical mistakes. Nevertheless, for the purpose of enhancing

the text's readability within this book, I have undertaken a process of 'naturalization'. During this process, certain excerpts have been refined by eliminating repetitions, interruptions, and instances where interviewees made self-corrections. Throughout this process, I took utmost care to preserve the original meaning and intention of the transcriptions (for a discussion, see Bucholtz, 2000).

References

Abbas, A.M., Hussein, R.S., Elsenity, M.A., Samaha, I.I, El Etriby, K.A., Abd El-Ghany, M.F. et al (2020) 'Higher clinical pregnancy rate with in-vitro fertilization versus intracytoplasmic sperm injection in treatment of non-male factor infertility: Systematic review and meta-analysis', *Journal of Gynecology Obstetrics and Human Reproduction*, 49(6): 101706.

Abramson, J.D., Rosenberg, H.G., Jewell, N. and Wight, J.M. (2013) 'Should people at low risk of cardiovascular disease take a statin?', *BMJ*, 347: f6123.

Abusief, M.E., Hornstein, M.D. and Jain, T. (2007) 'Assessment of United States fertility clinic websites according to the American Society for Reproductive Medicine (ASRM)/Society for Assisted Reproductive Technology (SART) guidelines', *Fertility and Sterility*, 87(1): 88–92.

Allan, H. (2007) 'Experiences of infertility: liminality and the role of the fertility clinic', *Nursing Inquiry*, 14(2): 132–9.

Armstrong, D. (2002) 'Clinical autonomy, individual and collective: the problem of changing doctors' behaviour', *Social Science & Medicine*, 55(10): 1771–7.

Armstrong, D. (2007) 'Professionalism, indeterminacy and the EBM project', *BioSocieties*, 2(1): 73–84.

Armstrong, S., Arroll, N., Cree, L.M., Jordan, V. and Farquhr, C. (2015) 'Time-lapse systems for embryo incubation and assessment in assisted reproduction', *Cochrane Database of Systematic Reviews*, (2). doi: 10.1002/14651858.CD011320.pub2.

Armstrong, S., Bhide, P., Jordan, V., Pacey, A., Marjoribanks, J. and Farquhr, C. (2019) 'Time-lapse systems for embryo incubation and assessment in assisted reproduction', *Cochrane Database of Systematic Reviews*, 5(5). doi: 10.1002/14651858.CD011320.pub4.

Armstrong, S.C., Vaughan, E., Lensen, S., Caughey, L., Farquhar, C.M. Pacey, A. et al (2023) 'Patient and professional perspectives about using in vitro fertilisation add-ons in the UK and Australia: a qualitative study', *BMJ Open*, 13(7): e069146.

ASA (2021) *Enforcement Notice: Advertising of Fertility Treatments.* Available from: https://www.asa.org.uk/static/cb6c3d82-5bfc-463a-9a9fac5cbfe6d831/f00db49e-0bde-4ccf-940807410ace3 5d9/Fertility-Treatments-Enforcement-Notice-FINALPDF.pdf (Accessed 24 August 2023).

Bahadur, G., Homburg, R., Bosmans, J.E., Huime, J.A., Hinstridge, P., Jayaprakasan, K. et al (2020) 'Observational retrospective study of UK national success, risks and costs for 319,105 IVF/ICSI and 30,669 IUI treatment cycles', *BMJ Open*, 10(3): e034566.

Bahadur, G., Woodward, B., Carr, M., Acharya, S., Muneer, A., Hornburg, R. et al (2021) 'IUI needs fairer appraisal to improve patient and stakeholder choices', *JBRA Assisted Reproduction*, 25(1): 162–4.

Balli, M., Cecchele, A., Pisaturo, V., Makieva, S., Carullo, G., Somigliana, E. et al (2022) 'Opportunities and limits of conventional IVF versus ICSI: it is time to come off the fence', *Journal of Clinical Medicine*, 11(19): 5722.

Barnett, E. (2023) 'When you're desperate to conceive, you'll pay anything, and IVF clinics are cashing in', 22 August. Available from: https://www.thetimes.co.uk/article/when-youre-desper ate-to-conceive-youll-pay-anything-and-ivf-clinics-are-cashing-in-v60djddcv (Accessed 22 August 2023).

Bartels, R.H.M.A. (2009) 'Evidence-based medicine: a marketing tool in spinal surgery', *Neurosurgery*, 64(6): E1206.

Becker, G. (1999) *Disrupted Lives: How People Create Meaning in a Chaotic World.* Berkeley, CA: University of California Press.

Becker, G. (2000) *The Elusive Embryo: How Women and Men Approach New Reproductive Technologies.* Berkeley, CA: University of California Press.

Bell, S.E. (2022) 'Symposium: risk, innovation and ignorance production in the field of reproductive biomedicine', *Reproductive Biomedicine & Society Online*, 14: 121–4.

BMJ Publishing Group (2009) 'The licensing of medicines in the UK', *Drug and Therapeutics Bulletin*, 47(4): 45–7.

Borsa, A., Bejarano, G., Ellen, M. and Buch J.D. (2023) 'Evaluating trends in private equity ownership and impacts on health outcomes, costs, and quality: systematic review', *BMJ*, 382: e075244.

Borup, M., Brown, N., Konrad, K. and Van Lente, H. (2006) 'The sociology of expectations in science and technology', *Technology Analysis & Strategic Management*, 18(3–4): 285–98.

Braakhekke, M., Mastenbroek, S., Moll, B.W. and van der Veen, F. (2017) 'Equipoise and the RCT', *Human Reproduction*, 32(2): 257–60.

Brandão, C., Rego, G., Duarte, I. and Nunes, R. (2013) 'Social responsibility: a new paradigm of hospital governance?', *Health Care Analysis*, 21(4): 390–402.

The British Academy (2019) *Principles for Purposeful Business*. Available from: https://www.thebritishacademy.ac.uk/publications/future-of-the-corporation-principles-for-purposeful-business/ (Accessed 30 January 2024).

Brown, N. (2003) 'Hope against hype – accountability in biopasts, presents and futures', *Science & Technology Studies*, 16(2): 3–21.

Bucholtz, M. (2000) 'The politics of transcription', *Journal of Pragmatics*, 32(10): 1439–65.

Callus, T. (2011) 'Ensuring operational compliance and ethical responsibility in the regulation of ART: the HFEA, past, present, and future', *Law, Innovation and Technology*, 3(1): 85–111.

Capalbo, A., Poli, M., Rienzi, L., Girardi, L., Patassini, C., Fabiani, M. et al (2021) 'Mosaic human preimplantation embryos and their developmental potential in a prospective, non-selection clinical trial', *American Journal of Human Genetics*, 108(12): 2238–47.

Cardona, B. (2008) ' "Healthy Ageing" policies and anti-ageing ideologies and practices: on the exercise of responsibility', *Medicine, Health Care and Philosophy*, 11(4): 475–83.

Carneiro, M.M., Koga, C.N., Mussi, M.C.L., Fradico, P.F. and Ferreira, M.C.F. (2023) 'Quality of information provided by Brazilian fertility clinic websites: compliance with Brazilian Medical Council (CFM) and American Society for Reproductive Medicine (ASRM) guidelines', *JBRA Assisted Reproduction*, 27(2): 169–73.

Carosso, A.R., van Eekelen, R., Revelli, A., Canosa, S., Mercaldo, N., Stura, I. et al (2021) 'Expectant management before in vitro fertilization in women aged 39 or above and unexplained infertility does not decrease live birth rates compared to immediate treatment', *Reproductive Sciences*, 29(4): 1232–40.

Carrick, M., Wilkinson, J., Polyakov, A., Kirkham, J. and Lensen, S. (2023) 'How do IVF patients interpret claims about fertility treatments? A randomised survey experiment', *Human Fertility*, 26(2): 347–54.

Carson, A., Webster, F., Polzer, J. and Bamford, S. (2021) 'The power of potential: assisted reproduction and the counterstories of women who discontinue fertility treatment', *Social Science & Medicine*, 282: 114153.

Cavaliere, G. (2023) 'Involuntary childlessness, suffering, and equality of resources: an argument for expanding state-funded fertility treatment provision', *The Journal of Medicine and Philosophy: A Forum for Bioethics and Philosophy of Medicine*, 48(4): 335–47.

Cavaliere, G. and Fletcher, J.R. (2022) 'Age-discriminated IVF access and evidence-based ageism: is there a better way?', *Science, Technology, & Human Values*, 47(5): 986–1010.

Chain, J. (2021) 'The HFEA 30 years on: what needs to change?', PET Annual Conference, 'Reproducing Regulation: Who Regulates Fertility and How?'

Chan, J.L., Schon, S.B., O'Neill, K.E. and Masson, P. (2014) 'An assessment of fertility clinic websites: room for quality improvement', *Fertility and Sterility*, 102(3): e243.

Chawla, S. (2023) 'IVF works for the lucky few. After a decade, I finally realised I wasn't one of them', *The Guardian*, 1 May. Available from: https://www.theguardian.com/commentisfree/2023/may/01/why-i-quit-ivf-fertility-business-money (Accessed 22 August 2023).

Chiware, T.M., Vermeulen, N., Blondeel, K., Farquharson, R., Kiarie, J., Lundin, K. et al (2021) 'IVF and other ART in low- and middle-income countries: a systematic landscape analysis', *Human Reproduction Update*, 27(2): 213–28.

Christianson, M.S., Goodman, L.R., Booth, R., Lindheim, S.R. and Azziz, R. (2021) 'Financial fluency: demystifying accounting and business planning for the reproductive medicine specialist', *Fertility and Sterility*, 115(1): 7–16.

Cirkovic, S., Wilkinson, J., Lensen, S., Jackson, E., Harper, J., Lindermann, K. and Costa-Font, J. (2023) 'Is the use of IVF add-on treatments driven by patients or clinics? Findings from a UK patient survey', *Human Fertility*, 26(2): 365–72.

Clarke, A.E. (2003) 'Situational analyses: grounded theory mapping after the postmodern turn', *Symbolic Interaction*, 26(4): 553–76.

CMA (Competition and Markets Authority) (2020) *The Competition and Markets Authority: Self-funded IVF Research*, qualitative research report. Available from: https://assets.publishing.service.gov.uk/media/5fa01b30e90e070420702a1b/IVF_Research_Final_Report.pdf (Accessed 20 April 2024).

CMA (2021a) *Guidance for Fertility Clinics on Consumer Law*. Available from: https://assets.publishing.service.gov.uk/media/60c233e6e90e07438ee574a7/Final_Guidance_for_Clinics__21.pdf (Accessed 20 April 2024).

CMA (2021b) *Fertility Treatment: A Guide to Your Consumer Rights*. Available from: https://www.gov.uk/government/publications/fertility-treatment-a-guide-to-your-consumer-rights/a-guide-to-your-consumer-rights (Accessed 20 April 2024).

CMA (2022a) *Consumer Law Compliance Review of Fertility Clinics. Findings Report*. Available from: https://assets.publishing.service.gov.uk/media/632d65af8fa8f51d1f83391a/A._Final_findings_report.pdf (Accessed 20 April 2024).

CMA (2022b) *Patients' Experiences of Buying Fertility Treatment. Qualitative Research Report*. Available from: https://assets.publishing.service.gov.uk/media/632c2fefe90e073721b08402/Consumer_research_report_160922.pdf (Accessed 14 July 2023).

Cochrane, A.L. (1972) *Effectiveness and Efficiency: Random Reflections on Health Services*, *eweb:2984*. Available from: https://repository.library.georgetown.edu/handle/10822/764041 (Accessed 21 July 2023).

Cohen, J. (2012) 'From Pythagoras and Aristotle to Boveri and Edwards: a history of clinical embryology and therapeutic IVF', in K. Coward and D. Wells (eds) *Textbook of Clinical Embryology*, Cambridge: Cambridge University Press, pp 177–92.

Cohen, J. and Alikani, M. (2013) 'Evidence-based medicine and its application in clinical preimplantation embryology', *Reproductive BioMedicine Online*, 27(5): 547–61.

Cohen, J., Alikani, M. and Franklin, S. (2015) 'The Oldham notebooks: a look back at one of the most remarkable scientific collaborations of the twentieth century', *Reproductive Biomedicine & Society Online*, 1(1): 1–2.

Cooke, I.D. (2016) 'Public and low-cost IVF', in S.D. Fleming and A.C. Varghese (eds) *Organization and Management of IVF Units: A Practical Guide for the Clinician*. Cham: Springer, pp 301–14.

Culley, L., Hudson, N. and Lohan, M. (2013) 'Where are all the men? The marginalization of men in social scientific research on infertility', *Reproductive BioMedicine Online*, 27(3): 225–35.

Cutting, R. (2012) 'Legislation in the United Kingdom', in Z.P. Nagy, A.C. Varghese and A. Agarwal (eds) *Practical Manual of In Vitro Fertilization: Advanced Methods and Novel Devices*. New York: Springer, pp 605–9.

Dadiya, J. (2022) 'Medical need and medicalisation in funding assisted reproduction: a right to health analysis', *Medical Law International*, 22(3): 249–74.

Daily Mail (2020) 'Scandal of the risky rip-off IVF "add-ons"'. Available from: https://www.dailymail.co.uk/news/article-9091465/Scandal-risky-rip-IVF-add-ons.html (Accessed 20 April 2024).

Daniluk, J.C. and Hurtig-Mitchell, J. (2003) 'Themes of hope and healing: infertile couples' experiences of adoption', *Journal of Counseling & Development*, 81(4): 389–99.

Datta, A., Campbell, S., Deval, B. with Nargund, G. (2015) 'Add-ons in IVF programme – hype or hope?', *Facts, Views & Vision in ObGyn*, 7(4): 241–50.

Datta, A.K., Maheshwari, A., Felix, N., Campbell, S. and Nargund, G. (2020) 'Mild versus conventional ovarian stimulation for IVF in poor responders: a systematic review and meta-analysis', *Reproductive BioMedicine Online*, 41(2): 225–38.

Deaton, A. and Cartwright, N. (2018) 'Understanding and misunderstanding randomized controlled trials', *Social Science & Medicine*, 210: 2–21.

Decoteau, C.L. and Underman, K. (2015) 'Adjudicating non-knowledge in the Omnibus Autism Proceedings', *Social Studies of Science*, 45(4): 471–500.

de Mouzon, J., Chambers, G.M., Zegers-Hochschild, F., Mansour, R., Ishihara, O., Banker, M. et al (2020) 'International committee for monitoring assisted reproductive technologies world report: assisted reproductive technology 2012', *Human Reproduction*, 35(8): 1900–13.

Dhont, M. (2013) 'Evidence-based reproductive medicine: a critical appraisal', *Facts, Views & Vision in ObGyn*, 5(3): 233–40.

Dhont, N. (2011) 'The Walking Egg non-profit organisation', *Facts, Views & Vision in ObGyn*, 3(4): 253–55.

Eclipse (2023) *Fertility & IVF Sector M&A Report*. Available from: https://www.eclipsecf.com/post/fertility-sector-m-a-rep ort (Accessed 20 April 2024).

Elder, K. and Johnson, M.H. (2015) 'The Oldham Notebooks: an analysis of the development of IVF 1969–1978. I. Introduction, materials and methods', *Reproductive Biomedicine & Society Online*, 1(1): 3–8.

Epstein, D. and ProPublica (2017) 'When evidence says no, but doctors say yes', *The Atlantic*, 22 February. Available from: https://www.theatlantic.com/health/archive/2017/02/ when-evidence-says-no-but-doctors-say-yes/517368/ (Accessed 20 July 2023).

ESHRE (2013) *IVF for 200 euro per cycle: first real-life proof of principle that IVF is feasible and effective for developing countries*, ScienceDaily. Available from: https://www.sciencedaily.com/releases/2013/07/ 130708103432.htm (Accessed 20 August 2023).

ESHRE Add-ons working group et al (2023) 'Good practice recommendations on add-ons in reproductive medicine', *Human Reproduction*, 38(11): 2062–104.

Evers, J.L.H. (Hans) (2017) 'Do we need an RCT for everything?', *Human Reproduction*, 32(3): 483–4.

Every-Palmer, S. and Howick, J. (2014) 'How evidence-based medicine is failing due to biased trials and selective publication', *Journal of Evaluation in Clinical Practice*, 20(6): 908–14.

Faircloth, C. and Gürtin, Z.B. (2018) 'Fertile connections: thinking across assisted reproductive technologies and parenting culture studies', *Sociology*, 52(5): 983–1000.

Fan, Y., Zhang, X., Hao, Z., Ding, H., Chen, Q. and Tian, L. (2017) 'Effectiveness of mild ovarian stimulation versus GnRH agonist protocol in women undergoing assisted reproductive technology: a meta-analysis', *Gynecological Endocrinology*, 33(10): 746–56.

Farquhar, C. and Marjoribanks, J. (2018) 'Assisted reproductive technology: an overview of Cochrane Reviews', *Cochrane Database of Systematic Reviews*, (8). doi: 10.1002/14651858.CD010537.pub5.

Fauser, B.C. (2019) 'Towards the global coverage of a unified registry of IVF outcomes', *Reproductive BioMedicine Online*, 38(2): 133–7.

Ferraretti, A.P., Gianaroli, L., Magli, M.C. and Devroey, P. (2015) 'Mild ovarian stimulation with clomiphene citrate launch is a realistic option for in vitro fertilization', *Fertility and Sterility*, 104(2): 333–8.

Fertility Network UK (2023) 'The impact of the cost-of-living crisis on fertility patients'. *Fertility Network Report*. Available from: https://fertilitynetworkuk.org/wp-content/uploads/2023/10/Fertility-Network-report-The-impact-of-the-cost-of-living-crisis-on-fertility-patients.pdf (Accessed 29 January 2024).

Fotaki, M. (2010) 'Patient choice and equity in the British National Health Service: towards developing an alternative framework', *Sociology of Health & Illness*, 32(6): 898–913.

Fotaki, M. (2013) 'Is patient choice the future of health care systems?', *International Journal of Health Policy and Management*, 1(2): 121–3.

Franklin, J.M., Platt, R., Dreyer, N.A., London, A.J., Simon, G.E., Watanabe, J.H. et al (2022) 'When can nonrandomized studies support valid inference regarding effectiveness or safety of new medical treatments?', *Clinical Pharmacology & Therapeutics*, 111(1): 108–15.

Franklin, S. (1997) *Embodied Progress: A Cultural Account of Assisted Conception*. London: Routledge.

Franklin, S. (2013) *Biological Relatives: IVF, Stem Cells, and the Future of Kinship*. Durham, NC: Duke University Press.

Franklin, S. (2019) 'A tale of two halves? IVF in the UK in the 1970s and 1980s', in V.C. Mackie, N. Marks and S. Ferber (eds) *The Reproductive Industry: Intimate Experiences and Global Processes*. Lanham, MD: Lexington Books, pp 15–30.

Frickel, S., Gibbon, S., Howard, J., Kempner, J., Ottinger, G. and Hess, D.L. (2010) 'Undone science: charting social movement and civil society challenges to research agenda setting', *Science, Technology, & Human Values*, 35(4): 444–73.

Gabe, J., Harley, K. and Calnan, M. (2015) 'Healthcare choice: discourses, perceptions, experiences and practices', *Current Sociology*, 63(5): 623–35.

Galiano, V., Orvieto, R., Machtinger, R. Nahum, R., Garzia, E., Sulpizio, P. et al (2021) ' "Add-ons" for assisted reproductive technology: do patients get honest information from fertility clinics' websites?', *Reproductive Sciences*, 28(12): 3466–72.

Geampana, A. and Perrotta, M. (2022) 'Accounting for complexity in healthcare innovation debates: professional views on the use of new IVF treatments', *Health*, 27(6): 907–923.

Geampana, A. and Perrotta, M. (2024) 'Using interview excerpts to facilitate focus group discussion', *Qualitative Research*. doi: 10.1177/14687941241234283.

Geampana, A., Perrotta, M., Giraud, A.S. and Hamper, J. (2018) 'The ethics of commercialization: a temporal analysis of newspaper coverage of IVF add-ons in the UK', *Human Reproduction*, 33: 279–80.

Geddes, L. (2022) 'From vaginal laser treatment to spa breaks – it's the great menopause gold rush', *The Guardian*, 26 January. Available from: https://www.theguardian.com/lifeandstyle/2022/jan/26/from-vaginal-laser-treatment-to-spa-breaks-its-the-great-menopause-gold-rush (Accessed 18 August 2023).

Gianaroli, L., Vitagliano, A., Ferraretti, A.P., Azzena, S., Terzuoli, G., Perruzza, D. et al (2022) 'IVF Lite: a smart IVF programme based on mild ovarian stimulation for good prognosis patients', *Reproductive BioMedicine Online*, 45(2): 256–63.

Gleicher, N., Kushnir, V.A. and Barad, D.H. (2019) 'Worldwide decline of IVF birth rates and its probable causes', *Human Reproduction Open*, 2019(3): hoz017.

Glenn, T.L., Kotlyar, A.M. and Seifer, D.B. (2021) 'The impact of intracytoplasmic sperm injection in non-male factor infertility: a critical review', *Journal of Clinical Medicine*, 10(12): 2616.

Goodman, L.K., Prentice, L.R., Chanati, R. and Farquhar, C. (2020) 'Reporting assisted reproductive technology success rates on Australian and New Zealand fertility clinic websites', *Australian and New Zealand Journal of Obstetrics and Gynaecology*, 60(1): 135–40.

Greener, I. (2009) 'Towards a history of choice in UK health policy', *Sociology of Health & Illness*, 31(3): 309–24.

Greenhalgh, T. (1997) *How to Read a Paper: The Basics of Evidence Based Medicine*. London: BMJ Publishing Group.

Greenhalgh, T. (2014) *How to Read a Paper: The Basics of Evidence-Based Medicine* (5th edn). Hoboken, NJ: John Wiley & Sons.

Greenhalgh, T. (2019) *How to Read a Paper: The Basics Of Evidence-Based Medicine and Healthcare* (6th edn). Hoboken, NJ: Wiley Blackwell.

Greenhalgh, T., Howick, J. and Maskrey, N. (2014) 'Evidence based medicine: a movement in crisis?', *BMJ*, 348: g3725.

Gross, M. and McGoey, L. (2015) *Routledge International Handbook of Ignorance Studies*. Abingdon: Routledge.

The Guardian (2018) 'How IVF became a licence to print money'. Available from: https://www.theguardian.com/lifeandstyle/2018/jun/18/how-ivf-became-a-licence-to-print-money (Accessed 20 April 2024).

Haddad, M., Stewart, J., Xie, P., Cheung, S., Trout, A., Keating, D. et al (2021) 'Thoughts on the popularity of ICSI', *Journal of Assisted Reproduction and Genetics*, 38(1): 101–23.

Haddiya, I., Janfi, T. and Guedira, M. (2020) 'Application of the concepts of social responsibility, sustainability, and ethics to healthcare organizations', *Risk Management and Healthcare Policy*, 13: 1029–33.

Haimes, E. (2013) 'Juggling on a rollercoaster? Gains, loss and uncertainties in IVF patients' accounts of volunteering for a U.K. "egg sharing for research" scheme', *Social Science & Medicine*, 86: 45–51.

Hamper, J. and Perrotta, M. (2023) 'Blurring the divide: navigating the public/private landscape of fertility treatment in the UK', *Health & Place*, 80: 102992.

Harbin Consensus Conference Workshop Group et al (2014) 'Improving the reporting of clinical trials of infertility treatments (IMPRINT): modifying the CONSORT statement', *Human Reproduction*, 29(10): 2075–82.

Harper, J., Magli, M.C., Lundin, K., Barratt, C.L. and Brison, D. (2012) 'When and how should new technology be introduced into the IVF laboratory?', *Human Reproduction*, 27(2): 303–13.

Harper, J., Jackson, E., Sermon, K., Aitken, R.J., Harbottle, S., Mocanu, E. et al (2017) 'Adjuncts in the IVF laboratory: where is the evidence for "add-on" interventions?', *Human Reproduction*, 32(3): 485–91.

Harvey, D. (2005) *A Brief History of Neoliberalism*. Oxford and New York: Oxford University Press.

Hawkins, J. (2013) 'Selling ART: an empirical assessment of advertising on fertility clinics' websites', *Indiana Law Journal*, 88(4): 1147–80.

Heneghan, C., Spencer, E.A., Bobrovitz, N., Collins, D.R.J., Nunan, D., Plüddemann, A. et al (2016) 'Lack of evidence for interventions offered in UK fertility centres', *BMJ*, 355: i6295.

Heng, B.C. (2007) 'Reluctance of medical professionals in adopting natural-cycle and minimal ovarian stimulation protocols in human clinical assisted reproduction', *Reproductive BioMedicine Online*, 15(1): 9–11.

Heymann, D., Vidal, L., Or, Y. and Shoham, Z. (2020) 'Hyaluronic acid in embryo transfer media for assisted reproductive technologies', *Cochrane Database of Systematic Reviews*, (9). doi: 10.1002/14651858.CD007421.pub4.

HFEA (Human Fertilisation and Embryology Authority) (2019) *Pilot National Fertility Patient Survey*. Available from: https://www.hfea.gov.uk/about-us/news-and-press-releases/2018/our-national-patient-survey-results/ (Accessed 20 April 2024).

HFEA (2021a) *Fertility Trend Report 2021*. Available from: https://www.hfea.gov.uk/about-us/publications/research-and-data/fertility-treatment-2021-preliminary-trends-and-figures (Accessed 12 July 2023).

HFEA (2021b) *State of the Fertility Sector 2020/21*. Available from: https://www.hfea.gov.uk/about-us/publications/research-and-data/state-of-the-fertility-sector-2020-2021/ (Accessed 20 April 2024).

HFEA (2022a) *National Patient Survey 2021*. Available from: https://www.hfea.gov.uk/about-us/publications/research-and-data/national-patient-survey-2021/#treatment-add-ons (Accessed 12 July 2023).

HFEA (2022b) *State of the Fertility Sector 2021/22*. Available from: https://www.hfea.gov.uk/about-us/publications/research-and-data/state-of-the-fertility-sector-2021-2022 (Accessed 1 August 2023).

HFEA (2023) *Fertility Treatment 2021: Preliminary Trends and Figures*. Available from: https://www.hfea.gov.uk/about-us/publications/research-and-data/fertility-treatment-2021-preliminary-trends-and-figures/ (Accessed 25 July 2023).

HFEA (nd) *Treatment Add-ons with Limited Evidence*. Available from: https://www.hfea.gov.uk/treatments/treatment-add-ons/ (Accessed 13 July 2023).

Hogarth, S. (2017) 'Valley of the unicorns: consumer genomics, venture capital and digital disruption', *New Genetics and Society*, 36(3): 250–72.

Homburg, R. (2022) 'IUI is a better alternative than IVF as the first-line treatment of unexplained infertility', *Reproductive BioMedicine Online*, 45(1): 1–3.

van Hoogenhuijze, N.E., Lahoz, N.E., Casarramona, G., Lensen, S., Farquhar, C., Kamath, M.S. et al (2023) 'Endometrial scratching in women undergoing IVF/ICSI: an individual participant data meta-analysis', *Human Reproduction Update*, 29(6): 721–40.

Houston, H.R. (2004) 'Other mothers: framing the cybernetic construction(s) of the postmodern family', *Consumption Markets & Culture*, 7(3): 191–209.

Iacoponi, O., van de Wiel, L., Wilkinson, J. and Harper, J.C. (2022) 'Passion, pressure and pragmatism: how fertility clinic medical directors view IVF add-ons', *Reproductive BioMedicine Online*, 45(1): 169–79.

IBISWorld (2022) *Fertility Clinics in the UK*. Available from: https://www.ibisworld.com/united-kingdom/market-research-reports/fertility-clinics-industry/ (Accessed 30 July 2023).

IFFS (2022) 'International Federation of Fertility Societies' Surveillance (IFFS) 2022: global trends in reproductive policy and practice, 9th edition', *Global Reproductive Health*, 7(3): e58–e58.

Inhorn, M.C. (2012) *Local Babies, Global Science: Gender, Religion and In Vitro Fertilization in Egypt*. Abingdon: Routledge.

Inhorn, M.C. (2015) *Cosmopolitan Conceptions: IVF Sojourns in Global Dubai*. Durham, NC: Duke University Press.

Ioannidis, J.P.A. (2016) 'Evidence-based medicine has been hijacked: a report to David Sackett', *Journal of Clinical Epidemiology*, 73: 82–6.

Jinkins, L.J., Parmar, A.D., Han, Y., Duncan, C.B., Sheffield, K.M., Brown, K.M. et al (2013) 'Current trends in preoperative biliary stenting in patients with pancreatic cancer', *Surgery*, 154(2): 179–89.

Johnson, M.H. (2011) 'Robert Edwards: the path to IVF', *Reproductive BioMedicine Online*, 23(2): 245–62.

Johnson, M.H. (2013) 'The early history of evidence-based reproductive medicine', *Reproductive BioMedicine Online*, 26(3): 201–9.

Johnson, M.H. (2019) 'A short history of in vitro fertilization (IVF)', *The International Journal of Developmental Biology*, 63(3–4–5): 83–92.

Johnson, M.H. and Elder, K. (2015) 'The Oldham Notebooks: an analysis of the development of IVF 1969–1978. VI. Sources of support and patterns of expenditure', *Reproductive Biomedicine & Society Online*, 1(1): 58–70.

Johnson, M.H., Franklin, S.B., Cottingham, M. and Hopwood, N. (2010) 'Why the Medical Research Council refused Robert Edwards and Patrick Steptoe support for research on human conception in 1971', *Human Reproduction*, 25(9): 2157–74.

Jureidini, J. and McHenry, L.B. (2020) *The Illusion of Evidence-Based Medicine: Exposing the Crisis of Credibility in Clinical Research* (illustrated edition). Adelaide: Wakefield Press.

Kadi, S. and Wiesing, U. (2015) 'Uninformed decisions? The online presentation of success and failure of IVF and related methods on German IVF centre websites', *Geburtshilfe und Frauenheilkunde*, 75(12): 1258–63.

Kamath, M.S., Mascarenhas, M., Franik, S., Liu, E. and Sunkara, S.K. (2019) 'Clinical adjuncts in in vitro fertilization: a growing list', *Fertility and Sterility*, 112(6): 978–986.

Kastenhofer, K. (2011) 'Risk assessment of emerging technologies and post-normal science', *Science, Technology, & Human Values*, 36(3): 307–33.

Keating, D., Cheung, S., Parrella, A., Xie, P., Rosenwaks, Z. and Palermo, G. (2019) 'ICSI from the beginning to where we are today: are we abusing ICSI?', *Global Reproductive Health*, 4(3): e35.

Keller, E., Botha, W. and Chambers, G.M. (2023) 'What features of fertility treatment do patients value? Price elasticity and willingness-to-pay values from a discrete choice experiment', *Applied Health Economics and Health Policy*, 21(1): 91–107.

Kieslinger, D.C., Vergouw, C.G., Ramos, L., Arends, B., Curfs, M.H.J.M., Slappenel, E. et al (2023) 'Clinical outcomes of uninterrupted embryo culture with or without time-lapse-based embryo selection versus interrupted standard culture (SelecTIMO): a three-armed, multicentre, double-blind, randomised controlled trial', *The Lancet*, 401(10386): 1438–46.

Kirkman, M. (2002) 'Moving on, living in the subjunctive mode: revising autobiographical narratives after infertility', *Meridian*, 18(2): 59–82.

Kirkman, M. (2003) 'Infertile women and the narrative work of mourning: barriers to the revision of autobiographical narratives of motherhood', *Narrative Inquiry*, 13(1): 243–62.

Knorr-Cetina, K.D. (1981) *The Manufacture of Knowledge: An Essay on the Constructivist and Contextual Nature of Science*. Amsterdam: Elsevier.

Knudsen, M. and Kishik, S. (2022) 'Multiplying ignorance, deferring action: dynamics in the communication of knowledge and non-knowledge', *Social Epistemology*, 36(3): 344–59.

Larocca, A. (2022) 'Welcome to the menopause gold rush', *The New York Times*, 21 December. Available from: https://www.nyti mes.com/2022/12/20/style/menopause-womens-health-goop. html (Accessed 18 August 2023).

Latour, B. (1987) *Science in Action: How to Follow Scientists and Engineers Through Society*. Cambridge, MA: Harvard University Press.

Latour, B. and Woolgar, S. (1979) *Laboratory Life: The Construction of Scientific Facts*. Princeton, NJ: Princeton University Press.

Lensen, S. (2020) *Special Collection: In vitro fertilisation – effectiveness of add-ons*. Available from: https://www.cochrane.org/news/spec ial-collection-vitro-fertilisation-effectiveness-add-ons (Accessed 21 July 2023).

Lensen, S., Sadler, L. and Farquhar, C. (2016) 'Endometrial scratching for subfertility: everyone's doing it', *Human Reproduction*, 31(6): 1241–4.

Lensen, S., Armstrong, S., Gibreel, A., Nastri, C.O., Raine-Fenning, N. and Martins, W.P. (2021a) 'Endometrial injury in women undergoing in vitro fertilisation (IVF)', *Cochrane Database of Systematic Reviews*, (6). doi: 10.1002/14651858. CD009517.pub4.

Lensen, S., Chen, S., Goodman, L. and Rombauts, L.J.F. (2021b) 'IVF add-ons in Australia and New Zealand: A systematic assessment of IVF clinic websites', *Australian and New Zealand Journal of Obstetrics and Gynaecology*, 61(3): 430–38.

Lensen, S., Hammarberg, K., Polyakov, A., Wilkinson, J., Whyte, S., Peate, M. et al (2021c) 'How common is add-on use and how do patients decide whether to use them? A national survey of IVF patients', *Human Reproduction*, 36(7): 1854–61.

Lensen, S., Armstrong, S., Vaughan, E., Caughey, L., Peate, M., Farquhar, C. et al (2023) ' "It all depends on why it's red": qualitative interviews exploring patient and professional views of a traffic light system for in vitro fertilisation add-ons', *Reproduction & Fertility*, 4(2): e220136.

Levine, H., Jørgensen, N., Martino-Andrade, A., Mendiola, J., Weksler-Derri, D., Mindlis, I. et al (2023) 'Temporal trends in sperm count: a systematic review and meta-regression analysis of samples collected globally in the 20th and 21st centuries', *Human Reproduction Update*, 29(2): 157–76.

Leyser-Whalen, O., Greil, A.L., McQuillan, J., Johnson, K.M. and Shreffler, K.M. (2018) ' "Just because a doctor says something, doesn't mean that [it] will happen": self-perception as having a Fertility Problem among Infertility Patients', *Sociology of Health & Illness*, 40(3): 445–62.

Luyten, J., Connolly, M.P., Verbeke, E., Buhler, K., Scotland, G., Lispi, M. et al (2022) 'Economic evaluation of Medically Assisted Reproduction: an educational overview of methods and applications for healthcare professionals', *Best Practice & Research Clinical Obstetrics & Gynaecology*, 85: 217–28.

Lyu, S.-R. (2015) 'Why arthroscopic partial meniscectomy?', *Annals of Translational Medicine*, 3(15): 217.

MacGregor, C., Petersen, A. and Parker, C. (2018) 'Hyping the market for "anti-ageing" in the news: from medical failure to success in self-transformation', *BioSocieties*, 13(1): 64–80.

Mackintosh, N. and Armstrong, N. (2020) 'Understanding and managing uncertainty in health care: revisiting and advancing sociological contributions', *Sociology of Health & Illness*, 42(S1): 1–20.

Macklon, N.S., Ahuja, K.K. and Fauser, B. (2019) 'Building an evidence base for IVF "add-ons"', *Reproductive BioMedicine Online*, 38(6): 853–6.

Maged, A.M., El-Mazny, A., Lasheen, Y. and El-Nassery, N. (2023) 'Endometrial scratch injury in infertile women undergoing in vitro fertilization cycles: a systematic review and meta-analysis', *Journal of International Medical Research*, 51(7): 03000605231175365.

Manning, S. (2018) *Couple who spent £20,000 on IVF treatment and £7,000 on add-ons sue*, *Mail Online*. Available from: https://www.dailymail.co.uk/news/article-6375673/Couple-spent-20-000-IVF-treatment-7-000-add-ons-UK-sue.html (Accessed 22 August 2023).

Marcus, H.J., Marcus, D.M. and Marcus, S.F. (2005) 'How do infertile couples choose their IVF centers? An internet-based survey', *Fertility and Sterility*, 83(3): 779–81.

Marriott, J.V., Stec, P., El-Toukhy, T., Khalaf, Y., Braude, P. and Coomarasamy, A. (2008) 'Infertility information on the World Wide Web: a cross-sectional survey of quality of infertility information on the internet in the UK', *Human Reproduction*, 23(7): 1520–5.

Masler, S. and Strickland, R.R. (2013) 'The role of management in an in vitro fertilization practice', *Seminars in Reproductive Medicine*, 31(3): 204–10.

Mastenbroek, S., De Wert, G. and Adashi, E.Y. (2021) 'The imperative of responsible innovation in reproductive medicine', *New England Journal of Medicine*, 385(22): 2096–100.

McGoey, L. (2014) *An Introduction to the Sociology of Ignorance: Essays on the Limits of Knowing*. Abingdon: Routledge.

McLaughlin, J.E., Knudtson, J.F., Schenken, R.S., Ketchum, N.S., Gelfond, J.A., Change, T.A. et al (2019) 'Business models and provider satisfaction in in vitro fertilization centers in the USA', *Journal of Assisted Reproduction and Genetics*, 36(2): 283–9.

McWilliams, A. and Siegel, D. (2001) 'Corporate social responsibility: a theory of the firm perspective', *Academy of Management Review*, 26(1): 117–27.

Meikle, J. (2004) 'Axe IVF watchdog, says fertility expert', *The Guardian*, 11 December. Available from: https://www.theguardian.com/society/2004/dec/11/health.medicineandhealth (Accessed 23 August 2023).

Morgan, D. (2004) 'Ethics, economics and the exotic: the early career of the HFEA', *Health Care Analysis*, 12(1): 7–26.

Morris, R. (2019) 'IVF and the "promise of happiness"', in V.C. Mackie, N. Marks and S. Ferber (eds) *The Reproductive Industry: Intimate Experiences and Global Processes*. Lanham, MD: Lexington Books, pp 97–107.

Moss, J. (ed.) (1998) *The later Foucault: Politics and Philosophy*. London and Thousand Oaks, CA: Sage Publications.

Mounce, G., Allan, H.T. and Carey, N. (2022) ' "Just have some IVF!": a longitudinal ethnographic study of couples' experiences of seeking fertility treatment', *Sociology of Health & Illness*, 44(2): 308–27.

Murad, M.H., Asi, N., Alsawas, M. and Alahdab, F. (2016) 'New evidence pyramid', *BMJ Evidence-Based Medicine*, 21(4): 125–7.

Murdoch, A. (2017) 'Should the HFEA be regulating the add-on treatments for IVF/ICSI in the UK?', *BJOG: An International Journal of Obstetrics & Gynaecology*, 124(12): 1849.

NICE (2013) 'Fertility problems: assessment and treatment'. London: NICE.

Nisker, J. (2009) 'Socially based discrimination against clinically appropriate care', *CMAJ: Canadian Medical Association Journal*, 181(10): 764.

Ombelet, W. (2013) 'The Walking Egg Project: universal access to infertility care – from dream to reality', *Facts, Views & Vision in ObGyn*, 5(2): 161–75.

Ombelet, W. (2014) 'Is global access to infertility care realistic? The Walking Egg Project', *Reproductive BioMedicine Online*, 28(3): 267–72.

Ombelet, W. and Campo, R. (2007) 'Affordable IVF for developing countries', *Reproductive BioMedicine Online*, 15(3): 257–65.

Ombelet, W. and Dhont, N. (2016) 'Emerging "cost-effective" treatments including low-cost IVF', in E.L. Stevenson and P.E. Hershberger (eds) *Fertility and Assisted Reproductive Technology (ART)* (7th edn). New York: Springer, pp 223–31.

Ombelet, W. and Goossens, J. (2016) 'The Walking Egg Project: how to start a TWE centre?', *Facts, Views & Vision in ObGyn*, 8(2): 119–24.

Ombelet, W. and Onofre, J. (2019) 'IVF in Africa: what is it all about?', *Facts, Views & Vision in ObGyn*, 11(1): 65–76.

Palomba, S., Vitagliano, A., Marci, R. and Caserta, D. (2023a) 'Endometrial scratching for improving endometrial receptivity: a critical review of old and new clinical evidence', *Reproductive Sciences*, 30(6): 1701–11.

Palomba, S., Carone, D., Vitagliano, A., Costanzi, F., Fracassi, A., Russo, T. et al (2023b) 'Fertility specialists' views, behavior, and attitudes towards the use of endometrial scratching in Italy', *BMC Women's Health*, 23(1): 397.

Panorama (2016) 'Inside Britain's Fertility Business', BBC One. Available from: https://www.bbc.co.uk/programmes/b084ngkd (Accessed 28 July 2023).

Patashnik, E.M., Gerber, A.S. and Dowling, C.M. (2020) *Unhealthy Politics: The Battle over Evidence-Based Medicine*. Princeton, NJ: Princeton University Press.

Patrizio, P., Albertini, D.F., Gleicher, N. and Caplan, A. (2022) 'The changing world of IVF: the pros and cons of new business models offering assisted reproductive technologies', *Journal of Assisted Reproduction and Genetics*, 39(2): 305–13.

Peddie, V.L., Van Teijlingen, E. and Bhattacharya, S. (2004) 'Ending in-vitro fertilization: women's perception's of decision making', *Human Fertility*, 7(1): 31–7.

Peddie, V.L., van Teijlingen, E. and Bhattacharya, S. (2005) 'A qualitative study of women's decision-making at the end of IVF treatment', *Human Reproduction*, 20(7): 1944–51.

Perrotta, M. (2013) 'Creating human life itself. the emerging meanings of reproductive cells among science, state and religion', *TECNOSCIENZA: Italian Journal of Science & Technology Studies*, 4(1): 7–22.

Perrotta, M. (forthcoming) 'Seeing through machines: algorithms and the pursuit of mechanical objectivity in embryo imaging', in S. Arnaldi, S. Crabu and A. Viteritti (eds) *Interfacing Bodies and Technologies: Practices, Imaginaries and Materiality*. Trieste: EUT.

REFERENCES

Perrotta, M. and Geampana, A. (2020) 'The trouble with IVF and randomised control trials: professional legitimation narratives on time-lapse imaging and evidence-informed care', *Social Science & Medicine*, 258: 113115.

Perrotta, M. and Geampana, A. (2021) 'Enacting evidence-based medicine in fertility care: tensions between commercialisation and knowledge standardisation', *Sociology of Health & Illness*, 43(9): 2015–30.

Perrotta, M. and Hamper, J. (2021) 'The crafting of hope: contextualising add-ons in the treatment trajectories of IVF patients', *Social Science & Medicine*, 287: 114317.

Perrotta, M. and Hamper, J. (2023) 'Patient informed choice in the age of evidence-based medicine: IVF patients' approaches to biomedical evidence and fertility treatment add-ons', *Sociology of Health & Illness*, 45(2): 225–41.

Perrotta, M. and Smietana, M. (2024a) 'Ensuring cost transparency and fully informed decision-making: addressing key needs, priorities and challenges'. Research Digest, No. 3. Queen Mary University of London, London. Available from: https://www.qmul.ac.uk/remaking-fertility/research-digests/ (Accesses 20 April 2024).

Perrotta, M. and Smietana, M. (2024b) 'Evidence gaps and information provision in fertility care: addressing key needs, priorities and challenges'. Research Digest, No. 1. Queen Mary University of London, London. Available from: https://www.qmul.ac.uk/remaking-fertility/research-digests/ (Accessed 20 Apri 2024).

Perrotta, M. and Smietana, M. (2024c) 'Information provision on UK fertility clinic websites: addressing key needs, priorities and challenges'. Research Digest, No. 2. Queen Mary University of London, London. Available from: https://www.qmul.ac.uk/remaking-fertility/research-digests/ (Accessed 20 Apri 2024).

Perrotta, M., Zampino, L., Geampana, A. and Bhide, P. (2024) 'Analysing adherence to guidelines for time-lapse imaging information on UK fertility clinic websites', *Human Fertility*. doi: 10.1080/14647273.2024.2346595.

Prasad, V.K. and Cifu, A.S. (2015) *Ending Medical Reversal: Improving Outcomes, Saving Lives*. Baltimore, MD: Johns Hopkins University Press.

Precedence Research (2021) *Fertility Market Report*. Available from: https://www.precedenceresearch.com/fertility-market (Accessed 20 April 2024).

Quaas, A.M. (2021) 'ICSI for non-male factor: do we practice what we preach?', *Journal of Assisted Reproduction and Genetics*, 38(1): 125–7.

Ray, A., Shah, A., Gudi, A. and Homberg, R. (2012) 'Unexplained infertility: an update and review of practice', *Reproductive BioMedicine Online*, 24(6): 591–602.

Repping, S. (2019) 'Evidence-based medicine and infertility treatment', *The Lancet*, 393(10170): 380–2.

Roberts, C. and Throsby, K. (2008) 'Paid to share: IVF patients, eggs and stem cell research', *Social Science & Medicine*, 66(1): 159–69.

Ruef, A. and Markard, J. (2010) 'What happens after a hype? How changing expectations affected innovation activities in the case of stationary fuel cells', *Technology Analysis & Strategic Management*, 22(3): 317–38.

Rumste, M.M. van, Evers, J.L. and Farquhar, C. (2011) 'Intra-cytoplasmic sperm injection versus conventional techniques for oocyte insemination during in vitro fertilisation in couples with non-male subfertility', *Cochrane Database of Systematic Reviews*, (2). doi: 10.1002/14651858.CD001301.

Sackett, D.L., Rosenberg, W.M.C., Haynes, R.B. and Richardson, W.S. (1996) 'Evidence based medicine: what it is and what it isn't', *BMJ*, 312(7023): 71–2.

Salecl, R. (2011) *The Tyranny of Choice*. London: Profile Books.

Sandelowski, M. (1987) 'The color gray: ambiguity and infertility', *Image: the Journal of Nursing Scholarship*, 19(2): 70–4.

Sauerbrun-Cutler, M.-T., Brown, E.C., Huber, W.J. and Frishman, G.N. (2021) 'Society for Assisted Reproductive Technology advertising guidelines: how are member clinics doing?', *Fertility and Sterility*, 115(1): 104–9.

Savulescu, J. (2011) 'The HFEA has restricted liberty without good cause', *The Guardian*, 7 February. Available from: https://www.theguardian.com/commentisfree/belief/2011/feb/07/hfea-reproductive-technology-research (Accessed 23 August 2023).

Schiller, R. (2019) 'The infertility premium: how big business exploits our deepest fears about pregnancy', *The Guardian*, 3 July. Available from: https://www.theguardian.com/society/shortcuts/2019/jul/03/why-a-vitamin-drip-wont-help-you-get-pregnant-the-rise-in-unproven-fertility-treatments (Accessed 20 August 2023).

Silva, S. and Machado, H. (2010) 'Uncertainty, risks and ethics in unsuccessful in vitro fertilisation treatment cycles', *Health, Risk & Society*, 12(6): 531–45.

Spar, D. (2006) *The Baby Business: How Money, Science, and Politics Drive the Commerce of Conception*. Harvard, MA: Harvard Business Review Press.

Spar, D. (2010) 'Free markets, free choice? A market approach to reproductive rights', in M. Goodwin (ed.) *Baby Markets: Money and the New Politics of Creating Families*. Cambridge and New York: Cambridge University Press, pp 177–90.

Spar, D. and Harrington, A.M. (2009) 'Building a better baby business', *Minnesota Journal of Science & Technology*, 10(1), 41–69.

Spencer, E.A., Mahtani, K.R., Goldacre, B. and Heneghan, C. (2016) 'Claims for fertility interventions: a systematic assessment of statements on UK fertility centre websites', *BMJ Open*, 6(11): e013940.

Stein, J. and Harper, J.C. (2021) 'Analysis of fertility clinic marketing of complementary therapy add-ons', *Reproductive Biomedicine & Society Online*, 13: 24–36.

Stevens, H. (2018) 'Evidence-based medicine from a social science perspective', *Australian Journal of General Practice*, 47(12): 889–92.

Stocking, K., Wilkinson, J., Lensen, S., Brison, D.R., Roberts, S.A. and Vail, A. (2019) 'Are interventions in reproductive medicine assessed for plausible and clinically relevant effects? A systematic review of power and precision in trials and meta-analyses', *Human Reproduction*, 34(4): 659–65.

Strathern, M. (1990) 'Enterprising kinship: consumer choice and the new reproductive technologies', *The Cambridge Journal of Anthropology*, 14(1): 1–12.

Swoboda, D. (2015) 'Frames of reference: marketing the practice and ethics of PGD on fertility clinic websites', in B.L. Perry (ed.) *Genetics, Health and Society*. Leeds: Emerald Group Publishing, pp 217–47.

Takhar, J. (2023) 'Communicative crises in the age of anxious reproduction and fertility preservation', *Consumption Markets & Culture*, 26(3): 210–16.

Takhar, J. and Houston, H.R. (2021) 'Forty years of assisted reproductive technologies (ARTs): the evolution of a marketplace icon', *Consumption Markets & Culture*, 24(5): 468–78.

Takhar, J. and Pemberton, K. (2020) 'Reproducing "rhetrickery" in online fertility marketing: harnessing the "rhetoric of the possible"', in E. Tissier-Desbordes and L.M. Visconti (eds) *Gender after Gender in Consumer Culture*. Abingdon: Routledge.

Tatsioni, A., Bonitsis, N.G. and Ioannidis, J.P.A. (2007) 'Persistence of contradicted claims in the literature', *JAMA*, 298(21): 2517–26.

Taussig, K.S., Hoeyer, K. and Helmreich, S. (2013) 'The anthropology of potentiality in biomedicine: an introduction to supplement 7', *Current Anthropology*, 54(S7): S3–S14.

Tehemar, S. (2015) 'Corporate social responsibility for healthcare organizations', in A. Örtenblad, C.A. Löfström and R. Sheaf (eds) *Management Innovations for Healthcare Organizations*. Abingdon: Routledge, pp 119–135.

Thompson, C. (2005) *Making Parents: The Ontological Choreography of Reproductive Technologies*. Cambridge, MA: MIT Press.

Throsby, K. (2010) '"Doing what comes naturally …": negotiating normality in accounts of IVF-failure', in L. Reed and P. Saukko (eds) *Governing the Female Body: Gender, Health, and Networks of Power*. Albany, NY: State University of New York Press, pp 233–52.

The Times (2016) 'Couples exploited by fertility clinics offering "add-ons"'. Available from: https://www.thetimes.co.uk/article/couples-exploited-by-fertility-clinics-offering-add-ons-wq8qzwdph (Accessed 20 April 2024).

Timmermans, S. and Angell, A. (2001) 'Evidence-based medicine, clinical uncertainty, and learning to doctor', *Journal of Health and Social Behavior*, 42(4): 342–59.

Timmermans, S. and Berg, M. (2003) *The Gold Standard: The Challenge Of Evidence-Based Medicine*. Philadelphia, PA: Temple University Press. Available from: https://www.jstor.org/stable/j.ctt14bt0qj (Accessed 20 July 2023).

Timmermans, S. and Kolker, E.S. (2004) 'Evidence-based medicine and the reconfiguration of medical knowledge', *Journal of Health and Social Behavior*, 45: 177–93.

Tippett, A. (2023) 'Reproductive rights where conditions apply: an analysis of discriminatory practice in funding criteria against would-be parents seeking funded fertility treatment in England', *Human Fertility*, 26(3): 483–93.

Tjørnhøj-Thomsen, T. (2005) 'Close encounters with infertility and procreative technology', in R. Jenkins, H. Jessen and V. Steffen, *Managing Uncertainty: Ethnographic Studies of Illness, Risk and the Struggle for Control*. Chicago: Museum Tusculanum Press, pp 71–91.

Topçu, S. and Maffi, I. (2022) 'Rethinking ignorance production in the field of reproductive biomedicine: an introduction', *Reproductive Biomedicine & Society Online*, 14: 216–21.

Trayer, J., Rowbotham, N.J., Boyle, R.J. and Smyth, A.R. (2022) 'Industry influence in healthcare harms patients: myth or maxim?', *Breathe*, 18(2): 220010.

Traynor, M. (2009) 'Indeterminacy and technicality revisited: how medicine and nursing have responded to the evidence based movement', *Sociology of Health & Illness*, 31(4): 494–507.

Van Blerkom, J., Ombelet, W., Klerkx, E., Janssen, M., Dhont, N., Nargund, G. et al (2014) 'First births with a simplified culture system for clinical IVF and embryo transfer', *Reproductive BioMedicine Online*, 28(3): 310–20.

Van Lente, H., Spitters, C. and Peine, A. (2013) 'Comparing technological hype cycles: Towards a theory', *Technological Forecasting and Social Change*, 80(8): 1615–28.

Vassena, R. (2023) 'How fast is too fast? Innovation in IVF and the burden of proof', *PET BioNews*, 3 July. Available from: https://www.progress.org.uk/how-fast-is-too-fast-innovation-in-ivf-and-the-burden-of-proof/ (Accessed 26 July 2023).

Vertommen, S., Pavone, V. and Nahman, M. (2022) 'Global fertility chains: an integrative political economy approach to understanding the reproductive bioeconomy', *Science, Technology, & Human Values*, 47(1): 112–45.

von Schondorf-Gleicher, A., Mochizuki, L., Orvieto, R., Patrizio, P., Caplan, A.S. and Gleicher, N. (2022) 'Revisiting selected ethical aspects of current clinical in vitro fertilization (IVF) practice', *Journal of Assisted Reproduction and Genetics*, 39(3): 591–604.

WHO (2017) *WHO Global Model Regulatory Framework for Medical Devices Including In Vitro Diagnostic Medical Devices*. Geneva: World Health Organization. Available from: https://apps.who.int/iris/handle/10665/255177 (Accessed 26 July 2023).

WHO (2023) *Infertility Prevalence Estimates, 1990–2021*. Available from: https://www.who.int/publications-detail-redirect/978920068315 (Accessed 4 July 2023).

van de Wiel, L. (2020a) *Freezing Fertility: Oocyte Cryopreservation and the Gender Politics of Aging*. New York: New York University Press. Available from: http://www.ncbi.nlm.nih.gov/books/NBK568235/ (Accessed 1 August 2023).

van de Wiel, L. (2020b) 'The speculative turn in IVF: egg freezing and the financialization of fertility', *New Genetics and Society*, 39(3): 306–26.

van de Wiel, L., Wilkinson, J., Athanasiou, P. and Harper, J. (2020) 'The prevalence, promotion and pricing of three IVF add-ons on fertility clinic websites', *Reproductive BioMedicine Online*, 41(5): 801–6.

Wilkinson, J., Roberts, S.A., Showell, M., Brison, D.R. and Vail, A. (2016) 'No common denominator: a review of outcome measures in IVF RCTs', *Human Reproduction*, 31(12): 2714–22.

Wilkinson, J., Vail, A. and Roberts, S.A. (2017) 'Direct-to-consumer advertising of success rates for medically assisted reproduction: a review of national clinic websites', *BMJ Open*, 7(1): e012218.

Wilkinson, J., Malpas, P., Hammarberg, K., Mahoney Tsidinos, P., Lensen, S., Jackson, E. et al (2019) 'Do à la carte menus serve infertility patients? The ethics and regulation of in vitro fertility add-ons', *Fertility and Sterility*, 112(6): 973–7.

Wilkinson, J., Brison, D.R., Duffy, J.M.N., Farquhar, C.M., Lensen, S. Mastenbroek, S. et al (2019) 'Don't abandon RCTs in IVF. We don't even understand them', *Human Reproduction*, 34(11): 2093–8.

Wilkinson, J., Bhattacharya, S., Duffy, J., Kamath, A., van Wely, M. and Farquhar, C.M. (2019) 'Reproductive medicine: still more ART than science?', *BJOG: An International Journal of Obstetrics & Gynaecology*, 126(2): 138–41.

Wilkinson, S. (2019) 'Selling hope: how wellness cashed in on fertility', *The Sunday Times*, 11 August. Available from: https://www.thetimes.co.uk/article/how-wellness-cashed-in-on-fertility-klcqcvfbj (Accessed 12 August 2023).

Wise, J. (2014) 'NICE calls for end to "postcode lottery" of fertility treatment', *BMJ*, 349: g6383.

Youssef, M.A.-F., van Wely, M., Mochtar, M., Fouda, U.M., Eldaly, A., El Abidin, E.Z. et al (2018) 'Low dosing of gonadotropins in in vitro fertilization cycles for women with poor ovarian reserve: systematic review and meta-analysis', *Fertility and Sterility*, 109(2): 289–301.

Index

References in **bold** type refer to tables. References to endnotes
show both the page number and the note number (145n6).